Victor Toyoji de Nozaki

Analysis of the Basic Sanitation Sector in Brazil

Victor Toyoji de Nozaki

Analysis of the Basic Sanitation Sector in Brazil

Analysis of services provided by the public and private sectors

ScienciaScripts

Imprint

Cover image: www.ingimage.com

This book is a translation from the original published under ISBN 978-3-330-77249-6.

Publisher:
Sciencia Scripts
is a trademark of
Dodo Books Indian Ocean Ltd. and OmniScriptum S.R.L publishing group

120 High Road, East Finchley, London, N2 9ED, United Kingdom
Str. Armeneasca 28/1, office 1, Chisinau MD-2012, Republic of Moldova, Europe
Managing Directors: Ieva Konstantinova, Victoria Ursu
info@omniscriptum.com

Printed at: see last page
ISBN: 978-620-8-64277-8

SUMMARY

To

My family

ACKNOWLEDGMENTS

After so much work, study, suffering, tiring classes, alternating moments of expectation, surprise, euphoria and discouragement, the end has come. Of course, all this work started back in undergrad, with the beginning of interest in some subjects or even the most interesting classes (often the classes were better than the content, as I had excellent teachers) and also with professional life itself, which led us to a particular area, in my case basic sanitation.

Before I begin my thanks, I would like to make a very important point to be discussed and rethought in the university's postgraduate programs, which is the exclusive dedication of students to the master's program, but without the counterpart of scholarships. It is very difficult for students to dedicate themselves exclusively to an activity without a financial contribution, not for profit, but for survival reasons. Quality comes after investment.

The first person I have to thank, without a doubt, is Prof. Dr. Rudinei Toneto Junior, for the wonderful teacher he is, the understanding advisor and especially the friend, not to mention, of course, his knowledge. Prof. Rudinei was the person who helped me discover the sanitation sector and showed me everything we could explore in it (during my undergraduate studies, when I was defining my monograph project at the end of 2003).

I would also like to thank a friend who always told me that sanitation is the most important thing in the world and said that sanitation doesn't make money, but we fall in love with it, Prof. Jair Bernardes, who always encouraged me to study sanitation, especially costs.

As for my family, I would like to thank my brother and mother who have always been by my side, and also Said who was patient and held things together for me while I worked on my dissertation. I would also like to thank my friends from college and the master's program, Carlos and Mazinha, with whom we shared good and bad times during our course.

It was not easy to complete this work, because working, studying, researching etc. takes time, patience and energy, but all that is in the past now, what is worthwhile in life are challenges.

SUMMARY

The aim of this dissertation is to analyze the current structure of the provision of basic sanitation services in Brazil and the institutional framework, seeking to identify the main factors that limit the expansion of services, the resumption of investments and the increase in efficiency, in an attempt to find possible solutions to overcome the problems of the sanitation sector, for example, greater private participation and/or ways of increasing efficiency in public provision.

It also seeks to present possible actions and measures to clean up the sector, as well as the needs it requires to universalize water supply and expand sewage collection and treatment in the country, either with greater private participation (with investment) or through policies and actions by the sector's managers, as well as by the government.

The analysis covered topics such as the low levels of coverage of water supply, sewage collection and treatment, the reasons and expectations for the sector, the entry of the private sector with investments in infrastructure, and especially in basic sanitation, as well as the legal instruments for this.

Based on SNIS data, it was possible to analyze the performance of sanitation service providers in southeastern Brazil (public and private), verifying the trend of each group of providers. Some experiences of private and public management were also selected, as well as a case of a publicly-owned company. The cities studied were Limeira, Jaboticabal, Serrana, Ribeirâo Preto and Campinas, all in the state of São Paulo.

The results show that private sanitation service providers have performed better than public ones in terms of administrative, financial, operational and technical issues. However, this does not mean that public providers are incapable of providing good services and that privatization is the only way to solve the problem, given the cases of Campinas, Jaboticabal and Serrana. However, it is necessary for the public authorities to adopt various measures to encourage greater investment in the sector. In addition, it was possible to see that the better performance of private providers is linked to an increase in tariffs, which causes a problem of access for the low-income population.

INTRODUCTION

There are many reasons to study the basic sanitation sector, not only in Brazil, but in a broader and more general way, since it is a sector that involves very peculiar characteristics, ranging from its form of provision, legislation, the existence of strong positive externalities in the area of health and the environment, among other characteristics and factors and, of course, the weight on the country's economic development and growth.

The basic sanitation sector in Brazil has had a low rate of investment for many years, due to macroeconomic problems, as well as government investment policy decisions. As a result, today the sector suffers the consequences of this low investment, which are: i) non-universal access to water, many Brazilians still do not have a water supply; ii) low sewage collection, even if we analyze only sewage collection, the Brazilian figures are extremely low; iii) practically zero sewage treatment, when you look at the treatment rate, the situation is critical for the country, and low investment is evident; iv) problems with waste[1] ; v) lack of urban drainage in cities[2] and vi) timid actions in the area of environmental education.

All these difficulties in the basic sanitation sector end up causing serious health problems for the population, the environment and consequently greater public spending. All of this is due to the greater likelihood and ease of contamination of the population by water-borne diseases and other diseases related to the lack of sanitation and infrastructure, such as the lack of proper disposal of waste in cities, creating "dumps"[3] , places where tons of waste are dumped and where animals share space with humans in the struggle for survival.

There are also problems such as the floods that occur in cities during the rainy season, caused by the pollution of rivers and streams, which causes great inconvenience and damage to residents and also to the public authorities, as well as the environmental issue that is extremely compromised, in view of environmental damage such as the pollution of streams, rivers, lagoons and areas contaminated with garbage (soil and even the water table).

In addition to all the quantitative and qualitative problems facing the basic sanitation sector in Brazil, what makes the problem worse is that this situation is not recent and has been going on for a long time, ever since Planasa, which was the largest investment and financing program for the basic sanitation sector in Brazil, Investments are very low and the results of the actions are not very sensible, which indicates low prospects for changes in the current situation, since the government continues to have the same difficulties as before, the lack of financial resources for investments, and the private sector is wary of the prospects for investments in the sector, in the current regulatory and legislative situation.

In an attempt to solve this problem, the federal government has promoted a series of actions to increase the

1 When it comes to solid waste, the country produces around 241,000 tons of garbage a day, 90,000 tons of which is household waste. Almost half of the waste collected is disposed of in dumps, without any kind of treatment. Another 45% is sent to controlled or sanitary landfills and 5% is treated in a plant.
2 The lack of urban planning, especially in urban drainage issues, leads to serious difficulties and problems such as flooding and inundation of cities, causing serious damage to residents.
3 The term garbage dump has come to be characterized as a place where all the city's waste is deposited in the open, without any form of treatment or proper packaging, and where various waste pickers look for materials to sell or even food for their survival.

resources available for the basic sanitation sector, such as the implementation of investment programs, centralization of the sector's actions in a single ministry (Ministry of Cities), as well as measures aimed at private participation in the sector.

To this end, the government promoted the Concessions Law, which was supposed to be a framework for private sector resources to be invested in public services, such as basic sanitation. Several other infrastructure sectors benefited from the law, but it was frustrating for the sanitation sector, which found itself on the margins of private investment.

This was followed by periods (1995 to 1998) in which investment increased (very timidly) and others of great retraction, motivated by government debt and credit restraint, promoted with the aim of achieving fiscal targets and high surpluses, plus the problems of legislation and regulation in the sector, which are very deficient. There is a great deal of hope in the attempt to boost the sector, and not just the sanitation sector, but the infrastructure sector as a whole, with the Public-Private Partnership (PPP), which is expected to channel a great deal of private resources into the public sector, especially by carrying out major works.

The aim of this dissertation is to analyze the current structure of the provision of basic sanitation services in Brazil and the institutional framework, seeking to identify the main factors that limit the expansion of services, the resumption of investments and the increase in efficiency, in an attempt to find possible solutions to overcome the problems of the sanitation sector, for example, greater private participation and/or ways of increasing efficiency in public provision.

To this end, the dissertation is divided into three chapters, in addition to the introduction and conclusion. The first chapter discusses the characteristics of the basic sanitation sector and the evolution of the sector in Brazil since the creation of the Banco Nacional da Habitaçâo - BNH and Planasa, as well as presenting various federal government programs and actions to finance the sanitation sector. In addition, the institutional framework for regulating the sector will be presented, outlining and discussing all the aspects and issues that make it difficult to attract investment, the justifications for investment and its importance to society.

The second chapter presents some international experiences in the provision of sanitation services, highlighting France and England, and defines the different models of service provision in Brazil, highlighting the differences between public and private management.

The third chapter compares the performance of public and private basic sanitation service providers in the south-east of Brazil, using information from the SNIS, and presents some cases of private and public management and one case of a public company. The cities studied were Limeira, Jaboticabal, Serrana, Ribeirâo Preto and Campinas, all in the state of São Paulo. The aim is to identify possibilities for resuming investment and improving service provision.

CHAPTER 1

Characteristics of the Sanitation Sector in Brazil

1.1 Basic Sanitation

Basic sanitation services, which in today's understanding include activities such as water supply, sewage collection and treatment, garbage, drainage and environmental education, are part of the so-called infrastructure services, along with other services such as: electricity (generation and transmission), telecommunications, public transportation, among others.

All of these infrastructure services in general (in most countries) are either provided by the public or regulated by the public, because all of their characteristics prevent the market from defining an efficient allocation or end up generating under-investment. This is why these services are provided by the public authorities or there is public intervention.

Thus, the government emerges as an element capable of intervening in the allocation of resources, acting in parallel to the private sector, seeking to establish the optimum production of goods and services that satisfy the needs of society[4] . In economic theory, the arguments for public intervention are:

I) Public Goods - The characteristic of a public good is that there is no rivalry or exclusion in its consumption .[5]

II) Externalities - This is an effect felt in a given sector, caused by an action in another carried out by people or even companies .[6]

III) Natural Monopoly - This is the characteristic of practically all public services, as they have a cost structure in which unit costs decline as the quantity produced increases, in addition to the existence of increasing returns 7 to scale.[7]

The set of infrastructure sectors are characterized in most cases by having the three characteristics mentioned above, even if in some cases they do not explicitly present all three, the characteristic of public good is present in all of them, with the characteristic of meritorious public good being the most common.

The basic sanitation sector has all three of the above-mentioned characteristics, i.e. it is a public good (even though it is meritorious), it produces externalities and it is classified as a natural monopoly. For a better understanding, we present explanations of each of its characteristics in order to be classified as a service that requires government intervention.

I) A public good, basic sanitation is a sector that has the characteristics of being a public good, because it is non-rivalrous, i.e. the use of water supply services and the collection and treatment of sewage by one

4 For more details, see Riani, 2002.
5 For more details, see Guena, 2000: 283.
6 For more details, see Guena, 2000: 259.
7 For more details, see Giambiagi, 2000: 26.

individual does not interfere with the consumption of another .[8]

II) Externality, basic sanitation is one of the infrastructure services that most clearly generate externalities, both in the areas of public health, the environment, the well-being of the population and economic growth itself. This is because investments to provide the population with better quality water, with efficient treatment, reduce the risk of transmitting diseases such as diarrhea and others related to water contamination, thus reducing the cost of curative public health.

Therefore, investing in basic sanitation will have a medium and long-term impact on reducing health costs, as the demand for medical services from sick people will be reduced. It should be noted that the definition of basic sanitation is very broad, and the current understanding of basic sanitation is changing to environmental sanitation, which is even more general, bringing together the issue of water, sewage, waste (household, hospital, construction, etc.), drainage, the environment and environmental education, i.e. a set of actions and activities that greatly interfere in the lives of citizens. And if all these items are properly provided for, the risks of diseases and parasites are greatly reduced.

It is known that most diseases linked to a lack of basic sanitation are linked to contaminated water[9] , so if there are policies aimed at providing the population with better quality water, correct sewage collection and especially a good hygiene education program, the promotion of reductions in the levels of diseases linked to sanitation tends to be reduced.

Mendonça and Motta (2005) used an econometric test to estimate the average cost of saving a life for each type of service: sanitation, education and public health, and came to important conclusions, such as the continued reduction of illiteracy is the cheapest alternative for lowering mortality rates. In addition, the authors confirmed the thesis that investment in sanitation is advantageous.

"Considering that access to sanitation services are preventive measures that, in addition to the positive externalities for the environment not accounted for here, avoid the risks and discomforts of disease, our results suggest that preventive sanitation actions, particularly water treatment, would be more economically justifiable for the continued reduction of infant mortality than defensive spending on health services." (MENDONÇA and MOTTA, 2005:12).[10]

As demonstrated in the work already mentioned and in the justifications presented, sanitation services are essential to life, generating great well-being for the population (generating many positive externalities), as studied by Person (2002), who verified the importance of sanitation for the well-being of the population. Data from the World Health Organization (WHO) indicates that every dollar invested in sanitation implies a reduction of approximately 4 to 5 dollars in medical costs.

---

8 It should be noted that if there is excessive consumption or even waste by some individuals, it can affect the availability of water for others, making the good non-rivalrous.

9 Diseases linked to transmission caused by poor sanitation are: cholera, gastrointestinal infections, typhoid fever, poliomyelitis, amoebiasis, schistosomiasis and shigellosis.

10 In this study, published by the Institute for Applied Economic Research - IPEA, there are other important conclusions, which refer to the importance of improving the coverage of sanitation services, as well as improving access to education and health services. The cheapest alternative to reducing mortality is to reduce illiteracy, as the two are closely linked.

Basic sanitation is also very important for the environment, in terms of protecting springs, bodies of water, streams, rivers, lakes, ponds, land, etc. With the new format of environmental sanitation, great importance is being attached to a global treatment of sanitation in all its nuances, i.e. investing in the exploitation of a good (water), but also promoting varied actions so that these sources are maintained and preserved.

It is in this sense that work on nature conservation, reforestation projects, policies and campaigns for the conscious consumption of water and, in particular, programs and policies for environmental education have emerged. For this reason, several federal government programs that transfer non-repayable funds from amendments or from the federal budget are characterized by the existence of obligatory counterparts from the municipalities that receive the funds.

And in the vast majority of cases, the required counterpart may be through the implementation of some kind of environmental education program or project, so many municipalities that receive these funds end up adopting this measure in order to meet the requirements of the agreement.[11]

As for the externalities generated by sanitation investments in terms of economic growth, we can point to the fact that investments in sanitation in most cases are related to major works, and that this ends up generating employment, wages and consequently growth in the economy.

III) Natural Monopoly is a characteristic of basic sanitation services ("network" characteristics), as they fit the characteristics cited by Giambiagi and Além (2000) of natural monopolies, such as increasing returns to scale, production costs decline as the quantity produced increases; depending on the sector, it is much more advantageous to have a single producer than several. In addition, with the existence of monopolies, the government can intervene to regulate the market if the monopoly is private, or even produce the good.

An example that illustrates the natural monopoly characteristic of the sanitation sector is the fact that it is economically unviable to compete with more than one service provider, because if company A has a water and sewage network already installed in the city, the cost for company B to enter the market is immense, in addition to the fact that company A already has privileged information about its consumers, their problems, their consumption profile, in other words, the information is asymmetrical compared to the incoming company. Also, physically, it is difficult to install two parallel water and sewage networks in the city, i.e. it is practically unfeasible from both a physical and economic point of view.

Basic sanitation services have some characteristics such as high implementation *costs (sunk costs),* because once the investments have been made it is practically impossible to use the assets for purposes other than basic sanitation[12] , they are essential services and also have a long maturation period and economies of scale.

All these characteristics therefore justify the need for the government to interfere in the sector, either by

---

11 Normally, all transfers of financial resources between a basic sanitation provider and the federal or state government are formalized by means of an agreement and a work plan, which set out all the rules, targets, requirements, conditions, amounts and timetable for the transfer of resources, as well as their purpose. Almost all the agreements are brokered by Caixa Econòmica Federal, which is responsible for monitoring and complying with the contractual conditions, through the institution's agencies and engineers.

12 For example, once a water or sewage network has been installed, if it is not used for any reason, it will be difficult for the investor to find another use for the asset or even to sell it.

providing services directly or indirectly, in order to regulate the activity carried out by another company. However, under-provision of infrastructure hampers economic growth. Various studies have highlighted the role of investment in infrastructure and show that there is a complementarity between public and private spending, where government investment in infrastructure encourages private investment, thus enabling greater productivity gains and economic growth in the country. The first study on the subject was by Aschauer (1989) who showed, using data for the American economy, that a 1% increase in the stock of public capital implied an increase in output of between 0.36 and 0.9%.

"Public capital has a greater impact on output than private capital. This could be due to a number of factors, including the externalities of public investment on private investment, or the fact that public capital is a precondition for private investment. Given these conclusions, the debate on competition between the public and private sectors for the resources available for investment could be replaced by complementarity between the two types of investment." (Toneto, 2004: 14)

1.2 History of sanitation in Brazil

In Brazil, the initial milestone of a policy by the federal government aimed at financing can be considered the creation and implementation of two instruments aimed exclusively at the sector: the National Sanitation Plan - Planasa and the Sanitation Financial System - SFS, managed by the now defunct National Housing Bank - BNH.

Both were created in the 1960s and had the administrative and managerial characteristics of the period, given the political situation at the time, i.e. the military government. As a result, many of the measures and policies adopted followed a centralizing line, aiming for a policy that was extremely focused on the federal government. As a result, the power of the central government was felt in the funding model and even in the management of the system in the country.

The National Housing Bank (BNH) was created in 1964 and had the mission of implementing an urban development policy. The SFS (Sanitation Financial System) was created within the BNH, which centralized and coordinated resources for the sector. Thus, there was a joint action with the state and municipal governments, with the counterpart that autonomous services or mixed economy companies were created.

Following the government's creation of Planasa, there was an incentive for municipalities to grant basic sanitation services to state sanitation companies (CESBs), as they were one of the only ones that had access to BNH loans, i.e. the central government used the power of granting resources to force municipalities to grant their basic sanitation services to state companies, promoting an attempt at centralization.

This characteristic once again demonstrates our country's form of government, which, according to Peixoto (1994), "in this task, backed by the authoritarianism of the federal government, the state governments and the CESBs used all the mechanisms of political pressure and real economic blackmail, bending the then innocent or subservient mayors and councillors in most Brazilian municipalities to their yokes and interests." This is so true that only 25% of municipalities have continued with autonomous services.

The peculiarity of the Plan, which made it one of the largest basic sanitation financing plans in the world, was the fact that it used resources from the FGTS (Guarantee Fund for Length of Service), established in 1967, with a monthly contribution of 8% of workers' monthly salaries.

The main guidelines were, according to the Ministry of Planning and Budget/Sepurb:

- Elimination of the deficit in water supply and sewage services, by extending services to all urban centers and social classes of the population, and subsequent maintenance of the balance between supply and demand, through a continuous process of planning and management; [13]

- Financial self-sufficiency of the sector, through the consolidation of state Water and Sewerage Funds (FAEs). The progressive financing of the FAEs would be carried out through transfers of funds from the FGTS and the budgets of the federal and student governments;

- Tariff policy to balance income and expenditure, while allowing cross-subsidization between consumers with higher and lower purchasing power within each company's jurisdiction;

- Development and consolidation of the State Sanitation Companies (CESBs), whose management should adopt a business philosophy;

- The federal government is in charge of the top management of the National Sanitation Policy, including a structure with the BNH at the top, initiatives taken by the various state governments to create CESBs and FAEs, execution of works and development of the sector through these CESBs, participation of municipal governments, granting the exploitation of their services to the CESBs and technical support to the BNH by technical bodies contracted by it;

- And global feasibility studies at state level and not at municipal level, or for each system.

The plan led to the creation of State Sanitation Companies (CESBs) in the states of the federation[14] . In order for them to operate in the states, it was necessary for the municipalities to grant concessions to the CESB in their state, given that the constitution established that the power to grant concessions rested with the municipalities; these concessions were signed for long periods of time.

Given the scale of the plan, with the large volume of resources, the favorable situation of the economy, the practice of cross-subsidies and loans at low rates, it was possible to expand services considerably. Some data shows this, according to Arretche, 1993: 2, for example, in 1980, the population served by Planasa's water supply was 50 million people (42% of the total population), and Planasa's sewage services covered around 17.5 million people. In 1990, when the total population was 146.8 million, Planasa's water services covered around 83 million people and sewage services 29 million people.

According to a survey carried out by Marcos T. Abicalil (1998), the basic sanitation sector had a major boost, growing 43% in water coverage and 122% in sewage collection between 1971 and the early 1980s.

For a better illustration, below is a table with information on the population served by Planasa's water

---

13 This is a fund in which employers contribute 8% of their employees' salaries every month.
14 A total of 27 companies were created.

supply.

Table 1.1

Evolution of the population covered by Planasa for water supply (1970-1991) - (in thousands of reais)

Years	North	North East	South East	South	Center West	Brazil
1970	455,6	2.582,2	7.238,8	1.289,3	314,7	11.880,8
1971	476,6	3.097,2	8.170,2	1.530,3	335,9	13.610,3
1972	499,1	3.805,3	8.522,1	2.259,6	354,5	15.440,7
1973	544,6	4.409,6	9.708,4	2.702,2	477,9	17.842,8
1974	604,8	5.037,0	10.206,6	3.441,2	1.206,8	20.496,6
1975	685,4	5.680,1	10.857,6	4.162,0	1.343,6	22.728,8
1976	938,9	6.149,4	15.699,9	4.659,3	1.630,1	29.347,7
1977	1.081,8	7.512,5	18.818,0	5.575,4	2.044,9	35.032,8
1978	1.171,1	8.288,5	20.782,1	6.285,6	2.260,6	38.788,0
1979	1.541,9	9.332,7	23.909,5	6.731,5	2.558,6	44.074,4
1980	1.836,7	10.567,0	28.632,0	7.852,1	2.969,8	49.568,5
1981	1.909,5	10.567,0	28.632,0	8.500,1	3.347,7	52.308,5
1982	2.112,4	11.494,5	30.389,6	8.500,1	3.596,6	56.093,3
1983	2.438,3	12.853,1	31.229,0	9.297,9	4.028,5	59.847,1
1984	2.647,6	13.831,7	32.755,0	9.801,9	4.380,1	63.416,4
1985	2.833,6	14.862,9	34.600,6	10.411,3	4.822,5	67.531,0
1986	3.029,8	16.214,4	33.738,6	11.070,1	5.299,1	69.352,2
1987	3.262,4	17.178,5	35.849,4	11.676,0	5.651,4	73.617,9
1988	3.531,6	18.202,8	38.124,2	12.642,2	5.933,2	78.434,2
1989	3.818,8	20.539,1	36.901,9	13.728,9	6.296,8	81.285,7
1990	4.116,6	21.122,8	37.737,6	13.535,6	6.392,8	82.905,6
1991	4.197,4	20.823,1	37.165,2	14.185,8	6.102,4	82.474,2

the vast majority in the most populous cities, and in the higher income segments, according to Arretche 1993: 02. It should be noted that in addition to these major differences, the plan gave greater preference to investments in water supply over sewage treatment. This is because water supply has a lower cost and gives service providers a faster financial return.

In a diagnosis of the sanitation sector carried out in 1994 as part of the Sector Modernization Project (PMSS), the investments made by PLANASA, PRONURB and with funds from the General Budget of the Union (OGU/TESOURO) were surveyed. Investments in water supply, by source and converted into dollars, are listed year by year in table 1.2, and investments in sewage in table 1.3.

Table 1.2

Investments made in water supply with Planasa/Pronurb and OGU resources - 1980/1993 (in US$)

	Source of funds		
Years	**Treasury**	**Planasa/Pronurb**	**Total**
1980	648.55	599,430.75	600,079.30
1981	13,156.50	851,458.37	864,614.87
1982	13,541.39	702,448.25	715,989.64
1983	6,063.32	448,684.78	454,748.10
1984	3,795.53	300,886.69	304,682.22
1985	5,315.07	442,313.37	447,628.44
1986	15,302.79	383,068.93	398,371.72
1987	15,764.68	478,099.05	493,863.73
1988	19,687.24	705,925.66	725,612.90
1989	33,053.00	476,935.19	509,988.19

1990	64,377.82	543,459.41	607,958.38
1991	60,377.82	411,570.58	471,948.40
1992	108,842.93	150,889.91	259,732.84
1993	174,290.48	95,362.15	269,652.63
TOTAL	**534,338.27**	**6,590,533.09**	**7,124,871.36**

Source: BGU and CEF/BNH

From the information in the tables you can see the size of the investments made by Planasa in relation to the treasury and the reasons why several authors classify the plan as one of the largest in the world. Investments in sewage are shown below.

Table 1.3

Investments made in sewage using Planasa/Pronurb and OGU resources - 1980/1993 (in US$)

	Source of funds		
Years	**Treasury**	**Planasa/Pronurb**	**Total**
1980	1,391.78	241,808.85	243,200.63
1981	1,668.55	342,023.40	343,691.95
1982	1,041.19	337,772.59	338,813.78
1983	546.02	155,958.37	156,531.39
1984	409.20	80,336.37	80,745.57
1985	161.68	166,596.75	166,758.43
1986	651.66	252,517.13	253,168.79
1987	4,199.14	401,418.12	495,617.26
1988	1,006.51	449,812.67	450,822.18
1989	4,275.62	282,144.28	286,419.90
1990	14,679.84	282,484.86	197,164.70
1991	24,377.84	118,881.47	143,259.31
1992	40,503.42	55,141.20	95,644.62
1993	67,380.03	34,276.72	101,656.75
TOTAL	**162,292.48**	**3,201,202.78**	**3,363,495.26**

Source: BGU and CEF/BNH

As can be seen from the last tables, and as already stated, Planasa's investments were geared towards prioritizing investments in the water supply system, leaving the sewage system with a much smaller volume of available resources.

Some of the plan's problems include its rigidity, being centralized, a characteristic of military governments, as well as not taking into account regional differences between the various corners of the country. The municipalities, which really had the power to grant funds, were unable to take advantage of the plan, because only the states could obtain funding, as has already been reported. Therefore, those that did not grant sanitation services to the CESBs had to pay for all the investments needed to provide the services out of their own resources.

Another important fact about the plan was that it practically aimed to invest in the construction and expansion of systems, without giving the necessary attention to the area of system operation; this ended up causing a deterioration in the sector, greatly increasing its inefficiency and level of losses, which today causes major losses for service providers, as well as being a justification for privatizing the sector.

In addition, in the 1980s, the world and, consequently, the country were going through a macroeconomic crisis, with a fragile economy due to high inflation. As a result, the sector was unable to charge consumers efficiently, which made it impossible for the sector to sustain itself, as the tariffs that existed were unrealistic, in addition to the fact that funding sources were ceasing.

Urban growth policy was often carried out in an unorganized way, causing the growth of slums on the outskirts of cities, making investment even more difficult. If it is extremely expensive financially to implement basic sanitation correctly with a well-formatted master plan, implementing basic sanitation in places without urban planning is much more expensive and technically difficult.

All of this ended up leading the CESBs into a great deal of debt, and over time they began to use outdated technologies, greatly increasing maintenance costs, as the whole system was geared towards expansion rather than maintenance and operation.

To make matters worse, there was a big drop in funding for the sanitation sector during the 1980s, due to the difficulty the country had in raising external funds, as well as the economic/fiscal crisis and lower productivity in the economy, which together affected investments in the sector, especially due to the lower volume of funds directed to the FGTS.

In 1986, the BNH was abolished and, with it, there was a change in the management structure of the basic sanitation sector, with management now carried out by the Sanitation Department of the Urban Policy Secretariat (DS/Sepurb) of the Ministry of Planning and Budget (1986), with Caixa Econômica Federal being responsible for managing the financial resources.

A provision was also introduced in the 1988 Constitution, which defined, in a not very clear way, that ownership of sanitation services was the responsibility of the municipalities, as it was a service of local interest, but in relation to metropolitan regions the picture remained undefined.

The various economic problems that affected the country and the world, such as the financial and economic crises, as well as various reasons, such as the interference of companies in the sanitation sector, led to falls in investment rates observed in the 1980s.

Since the 1990s, there has been greater institutionalization of water resource management, concessions and public service permits. In the period from 1974 to 1994, external loans for the basic sanitation sector were a great help in the process of investment in the sanitation sector. Detailed data by funding agency is shown in the next table.

Table 1.4

External loans applied to basic sanitation by agency and level of government - 1974/1994

Agency / Level of Government / Project Name	**Loan Amount (US$ million)**	**Date of Hire**
A - WORLD BANK - BIRD		
1, <u>Federal Government</u>		

Water Supply and Sanitation/MG (BNH)	36.0	06/17/1974
Water Supply and Sanitation/MG-II (BNH)	40.0	08/27/1976
Sanitation and Solid Waste/National (BNH)	110.0	03/10/1978
Water Supply and Sanitation for the	100.0	02/08/1979
Northeast (BNH)	130.0	04/14/1980
Water Supply and Sanitation for the Southern States (BNH)	139.0	07/188/1980
Water Supply and Sanitation/MG-III (BNH)	180.0	05/18/1981
Water Supply and Sanitation (BNH)	302.3	03/30/1983
Water Supply and Sanitation (BNH)	80.0	09/29/1988
. Rio Emergency (CEF)	80.0	12/30/1988
.PROSANEAR (CEF)	250.0	12/09/1992
. Modernization of the Sanitation Sector (MBES+BA, MS, SC)	1,447.3	
- Sub-total	280.0	02/16/1990
2. **State and municipal governments**	119.0	12/17/1992
Sanitation in São Paulo (SABESP)	117.0	12/17/1992
Water Quality Control (SP)	145.0	02/01/1993
Water Quality Control (PR)	661.0	
-Water Quality Control (MG)	2,108.3	
- Sub-total	(95.0)	
- TOTAL BIRD		
B - ITERAMERICAN BANK - BID		
1, **Federal Government**		
Sanitation Social Action Program (MBES)	350.0	04/29/1991
- Sub-total	350.0	
2, **State and Municipal Governments**		
Sanitation in the Metropolitan Area of São Paulo	63.0	06/09/1987
(SABESP)	77.5	02/09/1987
Microdrainage Program for the Municipality of São Paulo (City Hall)	100.0	09/28/1989
.Drinking Water and Sanitation System in Brasilia (DF)	145.0	01/26/1992

	199.2	12/09/1992
Drainage and Sanitation in Belém (PA)	450.0	12/17/1992
Sanitation in Fortaleza (CE)	350.0	02/09/1994
. Pollution control of the Tiete River (SABESP/SP)	1,384.7	
. Pollution control of Guanabara Bay	**1,734.7**	
- Sub-total		
- TOTAL BIRD		
C - KFW		
1, Federal Government		
2, State and Municipal Governments		
Basic Sanitation in Ceará	10.0	01/31/1990
Sanitation in Western Bahia	6.7	12/12/1983
Sanitation in Santa Catarina	6.7	06/29/1987
Flood Control on the Rio dos Sinos/RS	13.9	12/19/1984
- Sub-total	37.3	
- TOTAL KFV	**37.3**	
D - TOTAL		
1) BIRD	**2,108.3**	
Federal Government	1,447.3	
State and Municipal Governments	661.0	
2) IDB	**1,734.7**	
Federal Government	350.0	
State and Municipal Governments	1,384.7	
3) KFW	37.3	
State and Municipal Governments	37.3	
4) GRAND TOTAL	3,880.3	

Source: SEARB/SEPLAN-PR

With the National Privatization Program (PND) in 1995, the target was now the infrastructure sector; in addition, the Concessions Law was passed, which was the milestone for private sector participation in the infrastructure area, as there was now a legal basis for the concession of public services.[15]

The National Water Resources Management System was also created, thus institutionalizing the control and exploitation of water and basing the management of water resources on hydrographic basins.

Since the administration of Fernando Collor de Melo and the two administrations of Fernando Henrique Cardoso (1995 to 2002), the emphasis in the sanitation sector has been on modernization and the marginal expansion of service coverage. There were a number of institutional changes in terms of federal government management. Many initiatives were relatively successful in expanding services and modernizing the sector, which caused a structural change in the sector, ultimately hindering further progress in service and quality.

During this period, there were several federal government funding programs, with resources from the FGTS (Severance Indemnity Fund), the World Bank and the Inter-American Development Bank (IDB). The table below shows the main federal government programs:

Table 1.5

Federal government programs

Program	Period	Financing	Beneficiary/Developments
Pronurb	1990-1994	FGTS and counterpart	Urban population in general, with priority to low income
Pro-Sanitation	1995-	FGTS and counterpart	Predominantly areas with families earning up to 12 euros.
Pass	1996-	OGU and counterpart, IDB and Bird	Low-income population in municipalities with the highest concentration of poverty
Prosege	1992-1999	IDB and counterpart	Low-income population, favoring communities with incomes of up to 7 s.m.
Funasa-SB	-	OGU and counterpart	Technical and financial support for the development of actions based on epidemiological and social criteria
PMSS I	1992-2000	Bird and counterpart	Studies and technical assistance to states and municipalities at national level; investments in business modernization and increased coverage for Casan, Embasa and Sanesul
PMSS II	1998-2004	Bird and counterpart	Starts financing companies in the North, Northeast and Center-West and institutional development studies
PNCDA	1997-	OGU and counterpart	Rational use of water by sanitation service providers, suppliers and user segments

15 The PND was widely used in infrastructure sectors such as electricity and telecommunications.

FCP/SAN	1998-	FGTS, BNDES and counterpart	Private concessionaires in projects to extend coverage in areas with incomes of up to 12 s.m.
Propar	1998	BNDES	States, municipalities and concessionaires hiring consultants to make public-private partnerships viable
Prosab	1996-	Finep, CNPq, Capes	Research development in environmental sanitation technology

Source: Turolla (2002:15)

According to Turolla F. A. (2002), the federal government's programs can be divided into two sets of actions: one aimed at reducing socio-economic inequalities and favoring systems that lack economic and financial viability, and the other aimed at modernizing and institutionally developing sanitation systems.

With this in mind, the financing plans can be categorized according to the table below.

Table 1.6

Financing plans

Plan for Reducing Inequalities	Modernization Plan
Pronurb	PMSS I and II
Pro-Sanitation	PNCDA
Pass	FCP/SAN
Prosege	Propar
Funasa - SB	Prosab

Source: Turolla (2002), author's own elaboration

It is noteworthy that this period was marked by a more integrated sanitation policy with those of urban development, health and the environment. Funasa (the National Health Foundation) also contributed to the sanitation sector with investments and programs aimed at basic sanitation. As you can see, the entire sanitation policy has been left without a direct link to the Water Resources and Watershed Management System.

Even with the implementation of these policies, there was a retraction in investment in infrastructure between 1990 and 1998, as a percentage of GDP, which can be seen in the table below, and can, to a large extent, be explained by the currency devaluation of the period.

Table 1.7

Infrastructure investments from 1990 to 1998

Year	Investments (1) (US$ billion)	Share of GDP
1990	11,0	3,0
1991	9,4	2,5
1992	8,6	2,1
1993	8,6	1,9
1994	8,3	1,4
1995	9,3	1,5
1996	11,6	1,7
1997	13,7	1,6
1998	9,9	1,1

Total	90,4

***Source:* State companies and SEST. *Apud* Pêgo Filho et al (1999, p. 14).**

(1) Not including resources from the General Budget of the Union

According to Turolla (2002), an important difficulty was the focus of economic policy on resolving the effects of the crises in emerging countries, which affected Brazil, and the Brazilian crisis of 1999 itself.

This is because, due to the financial crises, credit to the public sector was restricted, drastically reducing the reserve of funds for the sanitation sector in the country. This can be seen from the data in Table 1.4.

Table 1.8

Average Investment per Period

Period	Average Annual Investment
1995-1998	R$1.3 Billion
1999	R$263 million
2000	R$21 million

Source: Turolla (2002), author's own elaboration

From all this, it can be concluded that the sanitation sector has been forgotten by the government to the detriment of other sectors.

In general terms, but precisely between 1995 and 1998, sanitation sector policy can be considered to have as its objective the decentralization of actions, greater participation and democratization of decisions and greater participation by the private sector. All this with the aim of increasing the supply of sanitation services in the country.

Some studies and researchers claim that the total amount of resources needed to achieve the objectives of the former Planasa, i.e. universalizing basic sanitation services, is in the order of five billion reais a year.[16] The following table shows some data on sanitation investment needs:

Table 1.9

Funding forecast for the investments needed to close the sanitation sector deficit in 2010 by source of funding

Sources	R$ (Billion/year)
Guarantee Fund for Length of Service - FGTS	2,5
Rates	1,5
General Budget of the Union - OGU	0,5
Others	0,2

Source: General Budget of the Union. Extracted from Justo (2004).

The figures, according to Montenegro (2000), can be reached by the government, and he also draws attention to the FGTS resources, which show differences between the amounts made available and those released, since in 1998 the amount made available was in the order of R$ 6 billion, with only R$ 1 million being lent out in the year.

16 Technical work developed by Montenegro (2000:04).

This demonstrates the government's policy of restricting credit and resources for the country's basic sanitation sector, thus imposing difficulties in contracting credit operations.

The sector is undergoing transformations and therefore has both legal and technical demands, such as: improving integration between all public sanitation policies, a specific regulatory framework for the sector (a topic that will be covered in more detail throughout the paper); reducing the level of bureaucracy so that small municipalities can obtain government funds - these municipalities are the ones that demand the most investment and have the worst administrative and technical structure; investing in the technical aspects of municipal service management, with the aim of improving quality and efficiency.

During the first two years of the Lula administration (2003 and 2004), the search for ways to invest in the sanitation sector has been ongoing. All responsibility now lies with the Ministry of Cities, which was created to promote a policy of urban infrastructure programs and actions.

The Public-Private Partnerships (PPPs) law was the solution found by the government to address the lack of financial resources in the country. The law was passed in 2004[17] , and was proposed as the way out for the country's economic growth. The law applies to direct and indirect public administration bodies controlled directly or indirectly by the Union, States, Federal District and Municipalities.

In short, the PPP is an administrative concession contract, through which the Administration delegates the execution of services from the public authorities to the private sector. The law also presents two types of concession: sponsored and administrative. The case of sponsored concessions refers to the concession of public services or works, when it involves, in addition to the tariff charged to users, a pecuniary consideration from the public partner to the private partner, i.e. it is a common concession where the state makes some form of consideration.

And administrative concessions are contracts for the provision of services of which the Public Administration is the direct or indirect user, i.e. the service is provided directly or indirectly to the public authority, such as the construction of hospitals and prisons.

The PPP law is an attempt by the federal government to create the ideal institutional environment for investments in infrastructure[18] , a lot of expectations were raised about the performance of the country's economy after this law, but the results are still practically non-existent.

"Although PPPs cannot, and should not, be considered as the solution to the country's infrastructure deficiencies, it would not be fair to underestimate the value that PPPs can bring to our country, especially if we consider the success stories in countries such as the United Kingdom, Chile and Portugal." (OLIVEIRA, 2005:4).

During the Lula government, there was an increase in contracts and financing for the sanitation sector,

17 Federal Law No. 11,079 of December 30, 2004.

18 The PPP Law is a law for the entire infrastructure sector, but it is perfectly suited to promoting investment in the basic sanitation sector. This demonstrates that the basic sanitation sector is still a sector on the margins of measures for its development.

according to data from the Secretariat for Government Communication and Strategic Management of the Presidency of the Republic, with R$1.7 billion and R$2.9 billion[19] for 2003 and 2004, respectively.

According to the IBGE Basic Sanitation Census, approximately 90% of households are served by a general water network and only 54% have sewage collection[20] . This characteristic of Brazilian basic sanitation, high levels of water supply and low levels of sewage collection and treatment, has a logic and reasons.

The main reason for the discrepancy between water and sewage is that a large part of the current results are due to the investments made by Planasa, and the plan's characteristic was to provide the population with water supply services, in other words, to universalize the services, as has already been seen. Thus, a large part of the plan's investments were channeled into water supply systems, as the idea was that all Brazilians would have the right to piped water.

In addition, the political effect of investments in water is much greater than in sewage. The effects of an investment in water supply in a locality that previously didn't have the services generates a feeling of gratitude for the politician who carried out the work, unlike if work were carried out to treat the local sewage. For this reason, managers and politicians prefer to invest in water supply rather than sewage collection and treatment.

Generally speaking, the sanitation sector is still in great need of investment, both in the area of water supply and sewage collection and treatment. There is also the fact that sanitation service providers do not have these resources for investment. The federal government has a great deal of investment power, which is the FGTS, but it is shared with other government investments, which means that there is a dispute over resources[21] , which also limits investment.

There are several solutions and possibilities for the sector, which we will present in the second capital, but there is a need for commitment and engagement from the government, legislators and civil society, as the legal, tax, administrative and even political landscape must change in order to implement new policies and form actions aimed at improving and expanding services.

1.3 Regulating the sector

The Basic Sanitation Sector, unlike other sectors such as electricity, telecommunications and even transport, has not been the target of a strong government movement towards privatization. Perhaps this is explained by the nature of the services, as they are extremely closely related to the public health of the population, being a classic government function, and also by their complexity, both technically and in terms of their legal structure.

Basic sanitation services have poor legislation, as well as some disagreements over responsibilities. For example, the 1988 Federal Constitution states that it is the Union's responsibility to set up the water resources

19 Amounts available.

20 Sewage collection does not mean sewage treatment, but only the removal of sewage through interceptors or outfalls, which in most untreated cities are taken to rivers, streams or the sea itself to be discharged *in natura.* If the rate of sewage treatment were taken into account, the percentage would be half that of sewage collection.

21 The FGTS funds are used by the government for investments in various areas and for various programs, so the allocation of the fund's resources for basic sanitation depends on political will and also on the "political strength" of the service providers in pressuring the government to release these resources.

management system, as well as to define the criteria for granting rights to its use, as well as other actions, but in its Article 30, item V, it does not clearly define the responsibility of municipalities for services:

"Article 30 - Municipalities are responsible for:

I - legislate on matters of local interest;

II - supplement federal and state legislation where applicable;

V - Organize and provide, directly or under concession or permission, public services of local interest, including public transport, which is essential."

As we can see, the justification for making basic sanitation services the responsibility of municipalities lies in the fact that they can and are considered essential services. There are also several other laws related to the sanitation sector, such as Law No. 8.987 of 1995 - the Concessions Law. This law is a milestone for privatization, as it established concession and permission regimes for the provision of public services, as provided for in article 175 of the Federal Constitution, i.e. it made possible and authorized the participation of private capital in activities that were previously prohibited.[22]

In 1995, Law No. 9.074 was also passed, establishing rules for granting and extending public service concessions and permits, which was important for providing investors with greater contractual guarantees.

It is also worth mentioning Law No. 9.433 of 1997 - the Water Resources Law, which created the National Water Resources Management System, and finally Law No. 8.666 of 1993, or the Bidding Law[23] , as it is known. This law is essential for the entire public and even private sector, as it is used both in the public management of activities and in the processes of concession (privatization) of public services.

This is because it applies to any public entity and is also the law to which the constitution refers for the processes of concession or granting of any public service, including basic sanitation services. The regulation of the sector is essential for the sector, but movements are needed in other areas such as the financial, credit, insurance and legal systems, among others.

"The regulation of the provision of services will be an element of any future institutional configuration to be created, but its range of power and its importance will be conditioned by the type of legal or political equation Iex-ante, of the tangle of vagueness that currently characterizes the basic sanitation sector." (ARAÙJO, 1999:73).

Several questions have been raised about this issue, especially when it comes to metropolitan regions. In these regions, the territorial limits of the municipalities are blurred, and there is a need for several systems to supply water to the region.

The metropolitan region of São Paulo, for example, has collection, treatment, adduction and reservoirs under the responsibility of SABESP (state ownership) and distribution under the competence of a municipal

22 These activities have been considered public in nature until now, such as energy, telephony, roads and even basic sanitation.
23 The bidding law will be explored further in this chapter.

company, demonstrating an arrangement that was promoted between Sabesp and the municipal company, which demonstrates the need for legal organization on the subject, to avoid possible problems, such as the refusal of one of the sides involved to renew contracts, settlement of values, etc.

In most of the country's Iiieiropoliian regions, the state companies obtained the concessions from the municipalities through contracts or because they were already operating there, with the exceptions of Betim - MG and Porto Alegre - RS.

For the basic sanitation sector to be interesting for private companies, it is necessary to clearly define the rules and the regulatory framework, especially on the issue of ownership of services, as the technical issue is also of fundamental importance.

In metropolitan areas, the ownership of services becomes much more important when you think about issues such as one municipality promoting a policy to reduce water consumption and its neighbor not. The whole policy has to be synchronized, which causes problems and makes any kind of action difficult.

In Brazil, the issue of water resources has been regulated by the Water Code, which has been in place for many years (since 1934). In 2000, the National Water Agency (ANA) was institutionalized. Its purpose is to implement water resources policy, thus integrating the national water resources management system, i.e. policies and initiatives aimed at water issues, as well as encouraging the creation of river basin committees.

In the country, all the existing regulatory agencies, such as ANATEL, ANEEL and ANP, were created as a result of the privatizations of their respective sectors and structured, and all the regulation of the sector was carried out with the process underway.

When we talk about regulation, we have to highlight, especially when analyzing other agencies in Brazil: the structure of contracts, legislation regarding concessions, tenders, etc., but also a legal and contractual way of demanding quality in the services provided to the population.

There is a trend towards reform, organization, debureaucratization and transparency throughout the public service, and the sanitation sector is no exception. However, there is nothing tangible in terms of regulating the sector.

There are many obstacles in the sanitation sector, such as the dispersion of power, one can imagine the number of existing municipalities, to know what will be the number of contracts, tenders, etc. that will be faced for a future privatization of the sector.

Even when it comes to metropolitan regions, there seems to be a consensus that the municipality really has the responsibility. The federal government tried, during the FHC administration, to pass a bill transferring responsibility for the provision of sanitation services in metropolitan regions to the state, but it was withdrawn from the Senate's agenda after numerous discussions.

Thus, another obstacle arises, which is the fact that if the government creates a supervisory body (agency) for the sector, it could be going against the Constitution, in view of the federative nature of the country, also violating ownership, since the responsibility for supervision should be the responsibility of the municipality that is granting the services.

What can be done with this is for the federal government to determine, in some general terms, and in a very broad way, because Brazil is immense and its diversity even greater, containing some basic key points, but nothing specific, because if it does this, it could interfere in the autonomy of the municipality, and it will hardly be able to foresee all the events and peculiarities of the local problems of each municipality, because this is the one responsible for overseeing compliance with the concession contract, and the one responsible for the services, being also the one closest to it.

In the event of privatization, the municipal authorities will be overburdened, as they will have to draw up all the bidding notices and concessions, set up a regulatory and supervisory agency, as well as foresee all the possible events during the years in which the services are concessioned. All of this is extremely difficult to draw up and can suffer from a serious problem, which is the influence of local politics.

It's clear that many municipalities have the technical capacity to do this, but the vast majority of small municipalities will suffer a lot to make the whole process viable, because if a municipality is interested in privatizing a certain sector, in most cases it's because it's inefficient. So you can be sure that there must be problems with the registration of networks, bad logistics in distribution and deficiencies in collection with incorrect registrations; all of this will be a major problem at the time of the bidding process.

No company is going to be interested in an object that is not clearly defined, so the municipality will have to do something about structuring it, at least in terms of registration. Drawing up a call for tenders and a contract is extremely important, as it will be the rule of the game, perhaps for many years to come, and not all municipalities are technically prepared or have the human resources for such an investment.

If that weren't enough, *there is* also the nefarious politics that exists in some parts of the country, where, in small municipalities, the power of politics is usually held by an oligarchy or economic group, which can distort the purpose of the process to benefit itself or its friends, given the small number of political forces or interests, which discourages any kind of competition.

There can also be breaches of contract on the part of the granting authority, because there are past political disagreements and the companies often have nothing to do with the case. One example is not allowing inflation to be passed on to the tariffs charged, etc.

There is also a problem that will begin to emerge now, which is the fact that the concessions granted to the State Sanitation Companies (CESBs) a long time ago are due to expire, and that many municipalities that have the services granted, in some cases, don't even have contracts in force, which will certainly cause problems for both sides.

Therefore, despite the importance and advantages of leaving the role of overseeing services to the municipality, due to its physical proximity, there are also counterweights that should be analyzed in more detail to avoid future problems with the concession of basic sanitation services.

To further complicate the sector, there is a fact that is often not mentioned, namely the relationship between basic sanitation and the public health, water resources and environment sectors; in fact, what should exist is integrated management.

As a result, in addition to a regulatory agency, to which information must be provided, the companies will also have to be accountable to and monitored by the Ministry of Health, in this case the Sanitary Surveillance, the State Departments that coordinate water issues, river basins (who knows in the near future, as this is what the ANA has been looking for) and environmental bodies; in addition, of course, to the most interested party in all of this, which is the population served.

All this indicates that the establishment of rules is necessary, but this will require a great deal of effort and difficulty, so that we have a sector with clear, well-defined rules and a safe environment for both investors and the public. This requires the commitment of the public authorities to promote the necessary changes and regulations, without forgetting to "listen" to the interests of the population, which is the most affected.

1.4 Responsible for monitoring and controlling sanitation

Water and sewage service providers are inspected and have to follow the rules of various agencies and even different ministries of the federal government. Since 2000, the Ministry of Health, through various entities such as Sanitary Surveillance and the National Health Foundation - FUNASA, has been implementing and inspecting the quality standards for water for human consumption, regulated by Ordinance No. 1,469 of December 20, 2000.

In this Ordinance, there are various regulations and stipulations as to who is responsible for monitoring it. It also establishes indicators for microbiological standards, turbidity, the presence of various chemical substances, radioactivity, among others, in other words, there is strict quality control of the potability of water, as well as various analyses that service providers must carry out in accordance with the Ordinance's sampling plan, many of which are highly complex and costly.

In addition to the Ministry of Health, service providers are also accountable to the Ministry of Cities and its other entities, since the Ministry of Cities took over this responsibility in 2003 (Lula government). It is still discreet and has very little presence, serving more as a centralizing body for the sector's financing policy, but it received a lot of criticism for its management in 2003, because the expected investments were not made.

It seems that the great hope of the Ministry of Cities is that the PPPs will actually make the necessary investments in infrastructure a reality, and thus also in the area of basic sanitation, since the agency has currently only invested and released funds for high-profile projects and works, leaving aside, at first, smaller works and projects, perhaps due to a political strategy decision.

In addition to these two entities, service providers also have other state inspection bodies, such as, in the case of the State of São Paulo, the Environmental Sanitation Technology Company - CETESB, the State Department for the Protection of Natural Resources - DPERN, the Department of Water and Electric Energy - DAEE, the Institute of Historical and Artistic Heritage - IPHAN, the Forestry and Environmental Police, the Public Prosecutor's Office, the State Court of Auditors and others.

In other words, there are a series of entities, bodies and regulations to be followed by service providers, in addition to providing prompt service to the population, which is the biggest stakeholder in the matter. Attending to all these entities, each with different requirements, with many norms and ordinances, imposes high costs.

One example is the need to comply with Ordinance 1469/2000, which requires many analyses throughout the year, but these analyses, due to their complexity, cannot be carried out by the service providers themselves (and there are regulations on this in the ordinance), which states that only accredited laboratories can carry out the analyses), so there is a need to hire a regulated laboratory to carry out the analyses, resulting in a cost which, according to SAAEJ, amounts to 40,000 reais a year, for a population of 70,000 inhabitants.

Another example that illustrates the difficulty posed by these various supervisory bodies is also provided by SAAE in Jaboticabal, which took more than two years to start building the city's sewage treatment plant, processing documents between these various bodies and entities, in addition to the high cost of approving the project. There was also the need to pay various fees to all the entities, including CETESB, in order for it to issue the Installation License (to start the work) and then the Operating License (to start the treatment plant); all of this imposes barriers for a small municipality to invest in basic sanitation.

1.5 Chapter Conclusions

As we have seen, sanitation services are part of the so-called urban infrastructure services that have demanded and continue to demand heavy investment for their universalization, i.e. for each and every Brazilian citizen to have access to their services, benefiting from all their positive characteristics.

Brazil has adopted various measures and plans to try to achieve its goal of providing all people with water and sewage services, but even with the world's largest financing plan using FGTS resources, Planasa has not met its service targets, and even 10 years after it was abolished, the situation is still critical, with an ever-increasing need for investment and, worse still, no prospect of a solution in the short term.

Because the service providers are not in a financial position to make the investments needed to meet demand on their own, their cost structure is very high, making it difficult to find the resources for investments, while the federal government also doesn't have enough resources to invest in the sector[24] , for various reasons, from restrictions to debt, the private sector does have the resources for investment, but the legal and administrative environment, the financial and insurance system in Brazil lacks development and action on the part of both the government and the private sector itself to create the ideal and real conditions to encourage investment.

The federal government has been trying to adopt measures to strengthen the country's institutional environment to receive these investments, such as the PPP Law, which, among other things, establishes general rules for bidding and contracting public-private partnerships within the public administration, In general terms, its objective is to seek alternatives for financing and managing public works and services, in a context of reducing the state's participation in the economy, but as already mentioned, it does not represent the end of the sector's

24 It is worth noting that, as has already been mentioned and discussed by several authors, the FGTS would be able to finance the sector in a way that would meet all the needs.

needs.

Finally, it should be emphasized that the path we can see from the government's actions in the country is to strengthen public sanitation service providers, through agreements and the release of funds that have never been released before, but also to encourage foreign capital investment in the sector, through modern legislation and other structural measures needed to create a favorable environment for investment.

CHAPTER 2

Possible solutions for the sanitation sector

As we have seen, the basic sanitation sector has a huge demand for investment and also the need to serve a large part of the population that lacks sanitation, either by providing good quality water or by collecting and treating sewage. Thus, the challenge for both government and society is enormous, coming up against a lack of financial resources for public entities, a lack of structure and preparation on the part of the current service providers, among other problems.

Brazil's case is no different from the situations experienced by other Latin American countries and Europe in the past. The solution adopted by countries such as France and England was to increase the participation of private capital and implement strong actions to monitor and encourage investment. In our case (Brazil), we see that there is the possibility of trying to promote a greater incentive for private capital to enter[25] or through the various programs of the Ministry of Cities and Health to create programs and actions to restructure the current providers of public basic sanitation services.

Let's first look at the solutions adopted by the British and French to solve their sanitation problems, remembering that the biggest companies in the world in this area are from these countries.

2.1 English case

The English process began with the centralization of basic sanitation service providers, so that they were unified into a few providers. This was promoted by the Water Act of 1973, which regionalized services based on the idea of river basins[26] . From 1989 onwards, privatization processes began to take place in England and Wales.

There are various justifications and objectives for the process, but in essence, the British government sought to provide sanitation service providers with better standards, as well as customer service, and the fact that the financial situation of the companies was delicate and there was no prospect of an improvement due to the country's economic situation, strongly influenced the process. Add to this the fact that the government in power in England at the time was the so-called "Liberals", who preached less state involvement in the economy in order to reduce inefficiencies and let the market organize itself and grow.

For Handmer and Jonson (2001:07), the privatization of sanitation in England and Wales was aimed at attracting private resources to make the necessary investments in order to bring these countries into line with the new European Union standards, which, according to Amparo and Calmom (2000:09), were estimated at US$ 40 billion.

The regulatory structure that was set up for the privatization of the sector in England was not simple, requiring a great deal of effort on the part of everyone involved in the process to implement it, as it is based on agents

25 It can be said that the government has tried to do this through laws and regulations that seek to organize and regulate the sector, making it more attractive to private capital.
26 A similar measure has been adopted in Brazil with hydrographic basins.

of quality economic regulation, in addition to the basic sanitation service providers themselves.

The government (secretary of state) was given responsibility for setting environmental standards and policies (water and sewage)[27] , and some inspectorates were given technical responsibility for checking quality. The *Office of Water Services* (OFWAT) was created to carry out economic regulation of the sector.

According to several authors, the British companies were sold for prices considered low, given the high cost the government had to pay to prepare them (including the sector) for the privatization process.

"The private sector benefited from the privatization by cancelling $12.5 billion in debts owed by the privatized companies to the government. This act aimed to increase the attractiveness of the process. In addition, the government injected US$ 3 billion into the privatized companies. As the stock market operation had a result of US$ 13.4 billion, there was a deficit of US$ 2.1 billion." (AMPARO and CALMOM, 2000:11).

Right after the privatization process, the companies increased their tariffs by approximately 5.5% above inflation[28] , after which the percentage went up to 1.5% above inflation.

This increase in tariffs, combined with the government's investments in canceling the companies' debts to the government, as well as a reduction in the companies' operating costs, culminated in a significant change in the economic and financial situation of the privatized companies. This made investments possible, as well as the payment of dividends to the companies' shareholders. In the period from 2000 to 2004 there was a reduction in tariffs and several questions from investors. It is worth noting that British companies are dominated by large groups of French companies.

Justo (2004: 30) presents a good characterization of the whole process of English privatization, based on a single issue, the financial one.

"The process of privatizing English sanitation has two "financial pillars". Firstly, the injection of funds to clean up the companies to be privatized. This solved the problem of high indebtedness.

The second pillar remained, namely adapting the companies' investment capacity to the interests of private capital. This issue was resolved by granting continuous increases in the real value of tariffs." (JUSTO, 2004:30).

The English process is in the final stages of adaptation and adjustment, but it serves as a good reference and privatization format for the basic sanitation sector. Ultimately, we can see that services have improved and efficiency has increased with private management.

2.2 French case

The French experience of privatization in the basic sanitation sector is the oldest. It is a decentralized model, because just as in Brazil, the responsibility for basic sanitation lies with the municipalities or communes. According to Parlatore (2000: 309), the first concessions of the *Compagnie Générale des Eaux* date from 1853

27 It should be noted that these standards are currently guided by the European Union.
28 According to BIRCHALL (2002:184), from 1990 to 1995.

and those of the *Lyonnaise des Eaux* from 1880. It is worth noting that France has the largest groups of companies investing in the sanitation sector[29] , and they are currently internationalized, with activities in several countries, including Brazil.

French municipalities have a great deal of autonomy and a certain amount of flexibility when it comes to selecting and contracting private services in the field of basic sanitation. The main characteristics of the types of contracts are set out in the table below:

Table 2.1

Public-private partnership models adopted in French sanitation

Models	
Risk-free models for the private sector	**Administration/Management Contracts *(la gérence)*** → the municipality hires a natural or legal person to manage the services, receiving a fixed fee that does not depend on their performance; → the role of the municipality: superior management of services, setting tariffs, meeting any deficits, absorbing surpluses and making investments; → function of the private operator: management of services and relations with users and third parties on behalf of the municipality. **Stakeholder Management Contracts *(régie intéressée)*** → similar to a traditional management contract, but with variable remuneration - varies according to contractually established performance indices **Lease contracts *(affermage)*** → the municipality hires an individual or legal entity, public or private, to manage and operate the services at its own risk; → investments are assumed by the municipality and the operator is paid directly by the user - tariffs established in the contract; → average term: 10 years; → returns to the city hall at the end of the period.
Models with risk for the private sector	**Concession contracts (concession)** → the public authority (granting authority) entrusts a legal entity (concessionaire) with the provision, financing and operation of the service at its own risk, under a long-term contract, and is paid directly by the users - tariffs defined in the contract; → Concessionaires are responsible for the necessary investments; → the works and facilities built are transferred to the municipality after the end of the period established in the contract;

29 The big French groups in the field of basic sanitation are: ONDEO-Suez, Vivendi and Bouyges. These groups invest not only in the sanitation sector, but in various other sectors and segments of the economy.

→ average term: 25 to 30 years.

Source: Saiani (2004:134)

According to Parlatore (2000: 209), the public management of services is done directly by the municipal administration or through decentralized entities, with their own legal personality, assets and budget, very similar to Brazilian municipal authorities. In the French case, the most important instrument in relations between private companies and municipalities are contracts, as there is no regulatory framework. In this way, contracts gain a lot of importance and strength in the process.

In France, there are also Basin Financial Agencies, which are responsible for managing the financial resources collected from water use charges, with the resources going towards water pollution projects. It is a similar structure to Brazil (in the case of the river basins, which also make investments and finance service providers and municipalities).

Turolla (2002:9) points out that one problem in France is the links between these large French groups and local government, contributing to corruption and fraud in privatization processes. However, in general, the French sector is very well developed and strong, providing the population with good services.

2.3 Public Power or Private Sector in Brazil

The majority of basic sanitation service providers in Brazil are public, with a few private[30] , Brazil is very similar to the French case, during the period of privatization of the French sector, in having decentralized services and municipal autonomy over water and sewage activity, the need for huge investments and the failure of the current forms of financing the sector. In recent years, we have seen two paths for the sector: many municipalities taking over water and sewage services for themselves[31] and the federal government promoting changes to the legal and institutional framework with a view to greater participation by private capital in the sector.

These two processes seem to be ambiguous, but as we shall see, they can coexist and bring good results for society, because society is only interested in having good quality sanitation services at fair prices, no matter who is responsible, but rather whether the good in question is meeting its needs.

The operation of Brazilian basic sanitation services by both public and private providers requires some institutional, legal, administrative and even government policy changes. For public providers, there is a need for a series of legal and political changes, especially when we refer to the debt limits imposed by the Fiscal Responsibility Law and also by the policies of governments to really invest in sanitation. When it comes to private providers, the changes that are needed are many, the need to change or create a legal system/regulatory framework for the sector, institutional and political issues are also necessary. Below we will see what these needs are for each of the types of management required for the country.

30 We will consider all existing types of private capital participation in the basic sanitation service to be private, as well as total participation (water and sewage) or only in one activity.

31 Most of the cases were municipalities in which the old concession contracts with the CESBs had expired or didn't even exist, but there are municipalities that tried privatization but decided to back out for some reason.

2.4 Forms of Local Public Provision of Basic Sanitation Services

2.4.1 Types of Provision

A very important factor that influences the performance of the basic sanitation service provider is the way in which it is provided. In Brazil, as *already* presented in the history of the sanitation sector, there are various forms of provision, which we can divide into three groups, as described below:

- Direct (Public Administration): Prefecture (Departments or other designation)
- Indirect (Public Administration): Autarchies and Public Companies;
- Indirect (Private Administration): Private Companies and Mixed Economy Companies.

Each of these forms of provision has some peculiar characteristics that differentiate them in various aspects, such as forms of administration, supervision, responsibilities, duties and other factors, all of which, of course, have a major influence on the performance of the provider.

The first group includes direct public administration, i.e. cases in which the municipality, through the town hall, is responsible for managing basic sanitation services. However, the services can also be carried out by a water and sewage department (or another department that assumes this responsibility)[32] of the municipal government.

In the second group are the municipalities that have opted to keep the basic sanitation services under their tutelage (in a public way), but in an indirect way, i.e. there is another entity that takes care of the management of the sanitation services, This happens in municipalities where an Autarchy or even a public company (a company owned by the municipality) is formed, which is responsible for managing the basic sanitation sector in the municipality.

And the last group is the case in which the municipality transfers the management of basic sanitation services to a company (private, mixed-capital company or other forms of private organization), leaving the municipality only with the role of overseeing the performance of the services, i.e. a regulator. [32]

The differences between the three types of service provision range from legal, administrative, operational and activity differences, since the legal format required for each of the entities is different, but it is logical that they should all fulfill the same role, i.e. to provide good quality basic sanitation services to the population in which they are based.

However, the entire administrative, accounting and management structure is different, and the way in which each of the three types is managed is very different and also has a varying degree of complexity. In this paper, we will divide the providers into public and private and carry out the analysis.

2.4.2 Characteristics of each type of provision

As we have already seen, each type of provision has its own characteristics and some peculiarities, such as the

---

32 The difference is merely one of nomenclature. Each prefecture has a different hierarchical structure, such as departments, sub-departments, inspectorates, secretariats, etc.

ways in which they are monitored, their relevant legislation, etc. Below we will present the main characteristics and differences of each of the forms of provision. Issues relating to the obligations of each of them, supervision, revenue, indebtedness and management will be addressed.

2.4.3 Obligations and supervision

All public entities are legally supervised by the legislative power, but it is technically assisted by the Courts of Auditors (Municipal, State or Union)[33] , whose competence, according to the Manual of the Court of Auditors (2006:06), is to act in accounting, financial, budgetary, operational and patrimonial supervision, in terms of legality, legitimacy, economicity, application of subsidies and waiver of revenue. The Court's jurisdiction extends to administrators and others responsible for public money, assets and values, as well as natural or legal persons who, by means of covenants, agreements, adjustments or other similar instruments, apply aid, subsidies or resources passed on by the public authorities.

In this case, in addition to the monitoring that is carried out by means of various accounting, management, financial, economic and even engineering information (in the case of construction, renovation or expansion works) that must be routinely forwarded to the headquarters of the courts, the courts of auditors also carry out annual *on-site* audits of public entities, be they city halls, public companies, municipalities or even entities maintained by the public authorities.

After the audit, and the analysis of the results of the monitoring of the information sent to the courts of auditors, opinions are issued on the audited entity, containing all the "infractions" committed by the public managers. Legal deadlines are given for justifications and defenses, when applicable. In the event that the entity is unable to justify itself, the report from the courts of auditors is sent to the legislature for it to judge and give a verdict on whether or not to approve the entity's accounts.

Currently, in addition to sending their reports to the legislative powers, the courts of auditors are also sending them to the Public Prosecutor's Office, which ends up opening proceedings to try the leaders involved, with possible punishments such as the return of funds, the revocation of political rights and even imprisonment.

In the case where the services are provided by the Direct Public Administration, i.e. through a department of the municipality (the first form of provision), the audit of the court of auditors is carried out together with the total accounts of the municipality, and it is treated as a department without any differentiation.

Of course, the court's audit will check whether the water and sewage department has been playing the role for which it is responsible, i.e. whether it is providing the population with adequate basic sanitation services, with water and sewage, how it is performing, how it is charging for its services, debtors, in other words, all the indicators of good management of the department, but the accounting, financial and even debt pieces are analyzed in the "cake" of the city hall .[34]

---

33 The Union is audited by the Federal Court of Auditors, and each state has its own court of auditors, which is responsible for the state and all its municipalities. However, in some states, such as São Paulo, Rio de Janeiro and others, in addition to the state courts, there is a special court for the capital.

34 This is because it is actually the Prefeitura (municipality) that is being analyzed.

When the service provider is the Indirect Public Administration (municipalities and public companies), i.e. the second form of provision, the entities, which in this case have legal, administrative and financial independence, are subject to exclusive supervision, unlike the previous case, but all points and questions are analyzed.

In this case, the court of auditors carries out a wide-ranging inspection of the entity, with a more detailed analysis of practically all the administrative points and factors involved such as:

- Meeting the objectives proposed for its existence (normally this inspection is based on the law that created the public authority or company, because in the body of the law, there is always a statement of reasons for its existence or creation), these objectives are in fact, whether or not the entity is meeting the needs of the population, with services in quantity and quality for all, of course within the limitations of the structure of the entity;

- Compliance with the Fiscal Responsibility Law and other existing legislation, i.e. checking that the entity has complied with all the rules and guidelines imposed by Federal Complementary Law No. 101, which sets spending limits and debt levels for public entities;

- In compliance with Law No. 4.320/64 - the Public Accounting Law, public entities have a very different accounting system from private companies, and this difference is based on this law, which determines, for example, how to draw up a budget, how to make records, how to make balance sheets and presentations, and other characteristics, such as commitments, appropriations, etc;

- Compliance with the Bidding Law - Law No. 8.666/93 - is the law that sets and determines the rules and procedures for purchases that a public entity must comply with. At first glance, it seems very bureaucratic and unnecessary, but it lives up to its name by trying to reduce fraud and corruption in the public sector;

- Compliance with the rules and instructions issued by the court of auditors, such as forms of contracting, drawing up contracts, publication of company acts, remuneration of its directors, etc;

- Compliance with good public sector management standards such as the entity's assets, debt collection, indebtedness, etc;

- In addition to various other existing standards.

In cases where the service provider is a private company, the court of auditors also inspects the entity, but the analysis is based on an analysis of the service concession contract, without administrative and financial interference on the provider, and of course whether the company is meeting the needs of the citizens.

It is worth pointing out that a concession can be canceled after the process has been analyzed by the court of auditors. Although the court has the role of being an auxiliary in the supervision of public entities, its opinions have been of fundamental importance and have also been the basis for actions by the public prosecutor's office to punish bad managers.

2.4.4 Revenue and Management

When it comes to a direct public administration entity, the collection of water and sewage tariffs or other types of services and related taxes, go into the so-called "cake" of the City Hall, so there is no link between the department's income and its respective expenses, that is, the provider (department) can collect much more than it actually spends, thus subsidizing another department or deficit area or the opposite can also occur.

There are cases in which the municipality carries out shared management between its departments, but this is not obligatory and is not a common feature of sanitation service providers, as it involves a certain degree of difficulty and bureaucracy to effectively carry out this monitoring.

A major problem or difficulty in the management of public services is the bidding process[35] , the purpose of which is to make contractual relations with the public authorities open (public), to ensure the right to participate and to avoid damage to public assets.

"Art. 1° This Law establishes general rules on bids and administrative contracts relating to works, services, including advertising, purchases, disposals and leases within the scope of the Powers of the Union, the States, the Federal District and the Municipalities.

Sole Paragraph. In addition to direct administration bodies, special funds, autarchies, public foundations, public companies, mixed-capital companies and other entities controlled directly or indirectly by the Union, States, Federal District and Municipalities are subject to the regime of this Law.

Works, services, including advertising, purchases, disposals, concessions, permits and leases of the Public Administration, when contracted with third parties, shall necessarily be preceded by a bidding process, except in the cases provided for in this Law.

Sole Paragraph. For the purposes of this Law, a contract is considered to be any agreement between bodies or entities of the Public Administration and private individuals, in which there is an agreement of will for the formation of a bond and the stipulation of reciprocal obligations, whatever the name used.

Art. 3 ° The bidding process is intended to ensure compliance with the constitutional principle of isonomy and to select the most advantageous proposal for the Administration and will be processed and judged in strict compliance with the basic principles of legality, impersonality, morality, equality, publicity, administrative probity, binding to the call for tenders, objective judgment and those that are correlated to them." (LAW N° 8.666/93)

The act of buying from a sanitation department (which is linked to the town hall) is not carried out in the department itself, but by the town hall, that is, by the body or department responsible for the purchasing and contracting sector, which in some cases ends up causing certain problems for the sanitation or water and sewage department.

Problems often occur due to the fact that purchasing is often a slow and bureaucratic process and that it takes a great deal of commitment on the part of the purchasing department to get it done in a short space of time.

35 Bidding is the act of buying and contracting that every public entity is obliged to follow, as defined by the Brazilian Constitution, bidding has its norms, characteristics, formats and other information, fixed by Federal Law No. 8.666, of June 21, 1993.

Thus, many purchases or contracts that are of fundamental importance for the development and performance of the water and sewage department and that necessarily go through the purchasing process (which is carried out by another department/sector) end up being delayed or even not carried out, due to the unwillingness of officials, political issues against the water and sewage department or other factors of the Brazilian bureaucracy of the municipalities, which is a case in point.

In addition, since the water and sewage department is directly linked to the city hall, it is subordinate to the city hall's wishes and action guidelines; if the water and sewage sector, i.e. basic sanitation, is not a government priority, the sector will certainly suffer from investment cuts, action reductions and cuts in purchases, contracts and tenders.

In the case of an indirect public administration, the entities have financial, administrative and legal independence; the revenue is collected by the service providers and it is the service providers who have sole control over the revenue, making the decision on how to manage their tariffs/revenue, as well as how to spend it.

This characteristic of indirect public service providers can be seen as a favorable differentiator for the management of services, because with greater administrative and financial independence of services and, consequently, resources, providers are not able to have their investment resources "cut off" by the municipality, as occurs in cases where the provider is directly linked to the municipality.

Even if the directors, superintendents or presidents of the municipalities or public companies are appointed by the mayor's office, which in a way makes this entity an extension of the mayor's office, because there is the political influence of the mayor's "boss", the result is better than in the previous case.

For the purposes of analysis and inspection, the court of auditors includes the results and data from the accounts of indirect entities (municipalities and public companies) in order to issue a financial and accounting result for the municipality, i.e. the town hall, because legally both a municipality and a public company form part of the municipality's assets, and are therefore its responsibility, and also the fact that these entities are replacing the municipality in the provision of services, which are inherent to public power.

Tendering is also compulsory, but it is carried out directly by the entity and not by the municipality, i.e. there is a body within the municipality or public company that is responsible for carrying out the tendering processes for purchases and contracts.

In this way, the problems seen earlier, when the service provider is directly linked to the town hall, cease to exist, because a purchasing or contracting process will certainly be "speeded up" by the municipality or company, if it is in the interest of the organization, in addition to the "pressure" and influence exerted by the organization's management, which has an interest in speeding up the process.

In the case of a private entity, revenues are not confused with those of the municipality, but rather, what is actually agreed between the municipality and the service provider, there is no obligation to tender for the company to contract with buy something, which drastically reduces bureaucracy.

2.4.5 Indebtedness

The issue of indebtedness must be addressed due to the fact that the sanitation sector suffers from a lack of financial resources, and there is a consensus that the sector alone cannot leverage financial resources for the necessary investments, so the search for resources through financing plays a fundamental role.

As a result, the degree of indebtedness is of great importance when it comes to the search for financial resources by service providers, in addition, of course, to the ability to pay, both of which are determined and set by the fiscal responsibility law.

In the case of direct administration, the indebtedness is actually the municipality's, i.e. the town hall and not the department, and in the case of indirect public administration, it is checked whether the entity is indebted or not. However, in practically all funding bodies or even for the federal government, the analysis must be carried out in conjunction with the municipality, and this is where the difficulty lies, because in practically all municipalities, sanitation service providers are in surplus and the municipality is in deficit, which ends up harming the autonomous provider.

In the table below we present summarized information on the topics discussed above:

Table 2.1

Comparing Public and Private Entities

Items	Public Entities	Private Entities
Inspection	- Legislative Branch - Court of Auditors	- Legislative Branch - Court of Auditors -External audits (depends on the legal form of the company) - no regulations exist
Purchasing procedures	- Law n° 8.666 - law on public tenders	legal, unless it is included in the concession or privatization
Accounting Procedures	- Law No. 4.320 - Public Accounting Law	- Follow the legislation in force, including the civil code
Indebtedness	- Fiscal responsibility law	- there is no legal regulation, unless it is included in the concession or privatization contract

Source: Tendering Laws, Public Accounting, Fiscal Responsibility and Constitution, own elaboration.

As seen in the table above, private managers or administrators may not have all the bureaucratic legislation that public ones have, but private ones, in some cases, have other requirements that end up being equivalent, and concession contracts also have the power over the rules to be complied with by managers and private entities.

2.5 Private sector participation

We have seen that the participation of private capital in the sanitation sector is still timid compared to other public service sectors, such as telecommunications, electricity and transportation (highways, ports, etc.), but for its participation, in addition to all the measures already adopted by the government and public authorities[36] , other measures are necessary.

In sectors such as telecommunications, during the privatization process carried out under the FHC government, much was questioned about the need to sell the federal government's state-owned companies, much was questioned about the final value of the sale of the companies and concessions, but what we have to point out is that the state didn't have the financial resources to invest in these areas, which require heavy investment, and also that their technological structure was far behind other companies in the same segment in the world.

After the whole period of transition from public to private management, and the "explosion" of the population's pent-up demand for such services, we saw an avalanche of complaints and questions, but after this turbulent period, we were able to see several benefits of competition[37] : a reduction in service prices, advances in the technology used, an improvement in the quality and quantity of the services provided, among other benefits. We can say that there has been a democratization of telecommunication services, as various strata of Brazilian society have started to use the services or have increased their consumption.

Toneto (2004: 41 and 42) says that the forms of private sector participation in the area of urban infrastructure are: (i) management and outsourcing contracts; (ii) *leasing*; (iii) concessions (with the models: BOT, BOOT and BOT) and (iv) privatizations. The complexity of participation varies, but can range from a simple case of outsourcing a service[38] to forms involving the participation of private capital in the management and operation of services/businesses.

Before going into the characteristics of each of the forms of private capital participation in the infrastructure sector, we need to highlight some basic points about public sector management. First of all, all the actions of public managers and their entities are governed by laws, the most important and supreme of which is the Brazilian Constitution, so several of the characteristics of the processes of private capital participation are motivated by this fact.

In this way, the Brazilian constitution determines points of public management such as: the need to hold a public tender for any type of contract between the government and another entity or company, the need to formalize contracts, the ways in which public servants are hired, among other issues.

For the purposes of this work, what is of most interest is the issue of tenders and contracts, because as

36 Some examples of government action to encourage private capital to participate in the infrastructure sector are: the privatization process, the concessions law and, lastly, the public-private partnership (PPPs) law, as well as other actions.
37 With the sale of state-owned companies, the change in the sector was basically twofold: private management of the business and a competitive market, because before the process, a single state-owned company operated in the sector and there was no competition.
38 It's a very common form of contracting today in practically every business sector.

determined by the constitution and the law on tenders itself (mentioned above), the public authorities, be they federal, state, municipal or entities linked to them, must carry out the tendering process for hiring any type of service or even for purchasing any type of product (whatever it may be), municipal or entities linked to them must carry out the bidding process for the contracting of any type of service or even for the acquisition of any type of product (whatever it may be)[39] , and furthermore, the legislation determines that in any transaction of purchase/sale or contracting the administration must carry out the formalization by means of an administrative contract.

Thus, in all forms of private capital participation in the sector, it is necessary to carry out a bidding process, as well as to formalize the agreements by means of an administrative contract. Of course, in some cases involving a complex, high-value contracting process or even a privatization process involving a public asset as important as basic sanitation, other formalities are required, such as: authorization from the legislature[40] , public hearings[41] , technical analyses, reports and a number of other items. Based on this information, we present below the characteristics of each of the models, according to Toneto (2004: 41 and 42):

- **Management and outsourcing contracts:** these are contracts between a public company and a private company which passes on the operation, management and maintenance of services to the latter for a certain period of time. There is also outsourcing, which takes place in certain specific activities of public management, with the aim of improving performance and reducing costs. Outsourcing usually takes place in the printing of water bills, "pothole repair" services (repair of the sidewalk removed for maintenance of the water supply or sewage collection system); reading of water meters, delivery of water bills, among other services and activities;
- ***Leasing:*** this involves leasing the ownership of the equipment for a certain period to the private sector, as well as the investments. In other words, ownership is public and management is private, and another characteristic of this format is that the private sector always assumes the operating risks. This format is widely used in the sanitation sector in France;
- **Concessions:** concessionaires are responsible for investments of all kinds and the replacement of fixed assets. The private sector operates the concession for a period of time, after which the assets revert to the public authorities. Concessions can take different forms with different characteristics, such as: payment of investments made by the private sector to the public, commercial and operational risk, which may or may not be transferred to the private sector?

---

39 Tendering is a bureaucratic process that is required in any case where the government contracts services or purchases products, regardless of the type of product/service or even its cost. However, in some cases, which are very well determined and proven, the legislation itself authorizes contracting or purchasing without carrying out this process, making it quicker to carry out and choose any supplier. Some cases that can be justified for waiving the call for tenders are: a state of calamity, lack of suppliers in the market, notorious specialization of the supplier, among other reasons (Law 8.666/93, art. 24).

40 The legislative power authorizes a concession, for example, by passing a law, authorized by the legislature itself or by the executive, in which it authorizes the public authority to grant a public service.

1 A public hearing is a process whereby the public administrator calls the whole population together to give notice of some act that the administration is going to carry out, for example, the concession of basic sanitation services. In this way, the entire population becomes aware of the act, as well as its justifications, and can offer suggestions, criticisms or even request that it not be carried out. This act serves as formal support from the population (those affected by the act) for the concession, if applicable.

BOT (Build, Operate and Transfer) - the private actor makes the investment, operates and transfers the assets to the public sector after the end of the concession contract;

BOOT (Build, Own, Operate and Transfer) - the feature that differentiates it from the basic model is that ownership must remain with the concessionaire, to serve as a guarantee;

BTO (Build, Transfer and Operate) - the concessionaire makes the investment (builds) but ownership is public.

- **Privatization:** All the assets that were public are transferred to the private actor who operates the services, in addition to the operation, ownership becomes private. As a result, the private actor assumes risks and various responsibilities, such as obtaining resources, investments, etc.

In relation to the participation of private capital in the public sector, there is an important discussion about the guarantees of operations and participation, in this sense there is *Project finance,* which is a financial engineering project. *Project finance* is characterized by the fact that a specific legal entity is formed to act on an investment or project. The guarantee for the operation is the business asset and not the project's corporations, as in other types of financing.

There are three financing formats in terms of collateral: one with a Treasury guarantee, so the risk is minimal; another with only the project's cash flow as collateral; and a third with a combination of the project's cash flow guarantee and an institutional guarantee.

In addition, with the Public-Private Partnerships law, there is a great expectation, at least on the part of the federal government, that the infrastructure sector (especially public works) will be leveraged with new investments from the private sector. As we have already seen, the government believes that the law is the way out of the country's economic growth. In essence, there is an attempt to get private capital to invest in activities/undertakings of public interest and with little or no economic return. In the case of PPPs, risks and investments are shared between private capital and the public sector.

PPPs, like other forms of private sector participation in the public sector, have important questions, which are in fact almost fundamental, in terms of guarantees for the operation and the business, in an attempt to reduce the risks involved. The big dilemma is how to attract private capital to a sector or business that is economically unviable or unprofitable, with high risks and low investment security, which require heavy investments. From this point of view, it is necessary for the public sector to participate in the venture in order to guarantee the safety of the operation, as well as to "show" investors that the government is working to ensure that the business performs well.

As we have seen, *there are* various ways in which private capital can participate in the public sector, especially in the urban infrastructure sector. Since the implementation of the so-called "privatization" process[42] , the

42 This was the period in which the government, using neo-liberal policies, opted for an ever smaller participation by the state in sectors that were typically dominated and monopolized by the public authorities. This happened with the sale of various government companies, as well as the concession of services that had previously been carried out exclusively by the public authorities.

government has adopted measures and actions aimed at attracting private investment. This is evident not only in actions for the basic sanitation sector, but for various other areas, since never before has the federal government worked so hard to improve Brazil's image abroad and increase the country's visibility abroad.

The government tries to create a suitable institutional environment for the participation of private capital, which is achieved through solid legislation that guarantees property rights, intellectual rights, etc. This is only possible if there is an excellent contractual structure in the country, involving not only the public authorities, but also the various players involved in contractual relationships.

In addition to the legal guarantees, there needs to be a political guarantee, i.e. that politicians won't interfere or alter any contract with the aim of electioneering or even out of private interest, so we can see that the demands go beyond the legal issue (laws) and a regulatory framework, Therefore, what private capital requires is a solid legal structure, but also a more subjective concept, which is respect for agreements by society, and a track record of "good antecedents".

All of this takes time to accomplish, in other words, demonstrating to investors that in this country there is respect for private capital, investors, property, in short, the laws not only enacted by the government, but also private or public administrative agreements and contracts.

Therefore, there are many measures to be adopted by the government to definitively attract the private sector to invest in areas such as basic sanitation, and the results are in most cases difficult to define. Fujiwara (2005:15) presents data showing that the reduction in mortality is associated with privatization[43] , and he also presents data showing that privatization has positive effects on water quality. Finally, evidence is presented that the privatization of basic sanitation services brings benefits to the less privileged population.

2.6 Characteristics of the Public Sector

As *already* discussed, the public sector maintains the vast majority of basic sanitation services in Brazil, either directly or indirectly. At this point, it's worth pointing out that all those involved are subject to the guidelines of the public authorities' legislation, as we've already seen (the bidding law, the fiscal responsibility law, the constitution, etc.), so the difficulties faced by public managers are manifold, and we'll go on to describe them.

Bureaucratic Management: public power is commonly associated with slow, inefficient work, without positive results and other similar adjectives, which we summarize as a bureaucrat. In short, bureaucracy doesn't necessarily have this popularly defined meaning. On the contrary, the bureaucracy defined by Max Weber had as its concept the idea of order, regulations to be followed, in other words, legislation that was widely known by everyone. Thus, we associate public power with bureaucratic management, and the reasons for this are as varied as possible.

One point is the existing legislation, i.e. a series of laws that the government must respect and comply with, which end up leaving it "plastered", these laws are administrative, financial and even operational. In the

43 In this study, using difference-in-differences estimators, there is evidence that the privatization of water and sewage services had a significant average effect of around 12% in reducing infant mortality in the municipalities of the states of São Paulo and Rio de Janeiro.

administrative area, there are laws such as the aforementioned law on tenders, the law on public accounting, the law on fiscal responsibility, organic laws, the law on public servants, and the constitution, among others. Thus, the difficulty in managing a public service is compounded by all this legislation, which often makes it difficult to buy or contract services that are largely emergency, when we think of basic sanitation services (an essential public good).

In many cases, instead of protecting the public purse and citizens' rights under the constitution, public tenders end up hampering public management to such an extent that they cause a lot of damage to the population, due to the slowness of purchasing/contracting processes, which may be emergency or even those that could reduce costs for the service provider or even improve the quality of services, which in some cases are delayed until the end of a bidding process .[44]

Public accounting also has a number of characteristics that make it an obstacle in the administrative process, because instead of the law following the evolution of accounting in the world, it works with methods and forms from 1964, as well as having an accounting language that is completely different from the usual language and form of the private sector. This is undoubtedly detrimental to the quality of services, because few professionals understand the subject, and so there is little choice in selecting good professionals, the techniques are old and there is no innovation or new forms of accounting management, which could contribute to better oversight and budget execution. And finally, the complex and different language ends up leaving public accounting data in a fog, making it difficult to disseminate information.

The government still has a serious problem with the way it hires its civil servants, because most of them are hired under the statutory system[45] , which brings with it a series of difficulties such as: job stability, various financial rights, little possibility of rehiring or reassigning a civil servant to another position, as well as, of course, the need to carry out long and time-consuming hiring processes, the public competition.

But the most serious and complicated point in the actions of the public authorities is related to the fiscal responsibility law, which was undoubtedly created with the aim of improving public services and punishing corrupt and inefficient administrators. However, in order to do so, it ends up creating serious barriers for various entities, such as the issue of indebtedness and payment capacity[46] , since the law sets maximum parameters, as well as indices to be met by managers. This ends up restricting credit to service providers, who are unable to overcome the limits imposed, thus preventing investments from being made and consequently improving the quality of life and economic growth, as has already been seen.

The public authorities have practically no federal, state, municipal or even private bodies to fund infrastructure projects, because it would be difficult for a private investor to be willing to put resources into a public entity

44 According to Federal Law No. 8.666/93, a bidding process is divided into four modalities: i) waiver of bidding; ii) invitation; iii) price-taking and iv) competition, each of these modalities has its own characteristics, and a minimum period for a process is 20 days, with no time limit for its conclusion, as legal actions can occur, a very common fact in important bidding processes.

45 Word that comes from statute, i.e. the law that regulates the work, rights, obligations and other provisions of public servants (employees of the public authorities).

46 For more information, see Federal Law No. 11.079 of December 30, 2004.

(for both technical and political reasons), so the only financiers are the federal government ministries (transfers made through agreements) or the state government, apart from these, the possibilities of extra resources are practically non-existent.

In addition, it is worth noting and stressing that the government's own resources are highly contested and are the target of political disputes, which in some cases end up prioritizing places where social needs are lower than in others that are not prioritized. In order to raise these funds from the government itself or, in some cases, from a private or international line of funding, the government has a very poor and unprepared technical staff. In other words, there are public servants who have no experience or technical ability to meet the demands and requests of these bodies.

One point that comes up is that it is common for public service providers to be short of technical staff, both administrative and operational. In addition, many public service providers don't have the technology or adequate structure to provide the service, leading to many complaints and a poor service. All of this was caused by managers who neglected to invest in human and technological capital, and now the backlog is too great, and investment is more than necessary.

In order to maintain services with the government, several changes are needed, such as: greater flexibility with regard to the restrictions on financing and indebtedness imposed by the fiscal responsibility law, opening up space for the contracting of debts by sanitation service providers public; real autonomy in relation to the municipality[47] , implementation by the federal government of a real policy for the sanitation sector, with clear and objective guidelines and goals, and with targeted rather than dispersed actions[48] . In other words, give special treatment to the sanitation sector, given all the benefits and importance it has, as has already been presented and discussed.

There are several possible paths, such as the case of Sabesp, a mixed-capital company whose largest shareholder is the Government of the State of São Paulo, which, under the rules and laws of the public sector, has used the Brazilian capital market to obtain funds to pay off debts and make investments. Like Sabesp, there are several companies, or even public entities, that could use the capital market as a way out of the lack of resources faced by the government, or even in view of the great bureaucracy of the current forms of financing, but investments in technical structure and professional training of the officials involved are necessary for operations in this direction.

2.7 Chapter Conclusions

In this capital we address issues such as the possible ways out for the basic sanitation sector to tackle its major problem of increasing investment and universalizing water supply and sewage collection and treatment services. The British and French cases show the state intervention and the solutions found by these countries

47 In cases where the service provider is a local authority or public company, the treatment given by the financing bodies is that of the entire municipality, which is detrimental to economic and financial performance, since in practically all municipalities the accounts of the town halls are in crisis. This prevents credit operations.

48 Currently, both the Ministry of Health and the Ministry of Cities are involved in sanitation and often both allocate funds to the same municipality, leaving aside another that has not received any investment.

to try to solve their problems. Of course, in neither case were the results exceptional or problem-free in terms of implementation and operation, but in both cases there was an increase in investment and an improvement in services.

In Brazil, in view of the way it is structured, the possible way out for the sector is to invest in the public sector (public providers) combined with legal and administrative changes to make it possible for service providers to act more flexibly, especially in terms of financing.

There is also the possibility of increasing the participation of private capital in the sector, some of which the government has been promoting, such as the concession law, the PPP law and others. However, there are still other legal, social and security needs to encourage the participation of private capital as a financier of sanitation services. Both PPPs and *Project Finance* are possible solutions to be adopted in the case of using private capital in the sector.

However, what has not been discussed in depth so far is the performance of both public and private providers, so in the next capital we will present data on the performance of public and private providers in the south-east of the country, as well as some cases of private concessions and other successful public managements.

CHAPTER 3

Quality Indicators for Water and Sewerage Service Providers

With the aim of measuring and trying to ascertain whether there is a tendency for the form of management of the sanitation service provider (public or private) to influence its performance. Data was collected from SNIS among public and private providers in the southeast of the country.

The indicators for the water and sewage service providers studied were obtained from SNIS (2001), so a short description of the system is necessary, as well as how the information was obtained, before presenting the indicators and service providers and their performance.

3.1 National Sanitation Information System - SNIS

The SNIS was conceived by the Federal Government in 1995, as part of the Sanitation Sector Modernization Programme - PMSS, linked to the Special Secretariat for Urban Development of the Presidency of the Republic, and developed with the support of the Institute for Applied Economic Research - IPEA.

Under the current Lula government, the system is being managed by the National Secretariat for Environmental Sanitation of the Ministry of Cities.

The idea behind the SNIS is that, when working with information in an organized, objective and targeted way, it is possible to work and manage a sector, in this case the basic sanitation sector, with greater subsidies for policy formulation and action planning, with a view to maximizing investments, thus increasing the efficiency and effectiveness of actions.

The SNIS consists of a federally managed database containing information on public water and sewage service providers in various areas, such as operations, management and finance, as well as some information on the quality of services throughout the country, which has been updated annually since 1995.

For the federal government, the data is used to assist in the planning and execution of public policies, guiding the application of resources, action strategies, monitoring the benefits and effects of investments and programs, as well as evaluating the performance of service providers.

Over time, the SNIS has demonstrated its importance as a tool for planning actions, not only by the government, but also by investors and financial, educational and research institutions.

3.2 Collecting Information for SNIS

The SNIS sample is made up of all regional and micro-regional service providers and a subset of local providers. Over time, we have sought to expand the sample, taking into account the following aspects:

- reach all regions of the country in a balanced way;
- improving the balance between municipalities according to their population;
- increase the participation of small entities.

To compose the sample and obtain the data, after selecting the water and sewage service providers, according to the aspects already considered, they are sent the Information Collection Program on CD-ROM, which must be filled out in a self-explanatory and easy-to-use way. All SNIS information is collected and processed in a specific program developed for this purpose and then entered into a database.

However, the sample to be analyzed is very diverse, including service providers from the most varied regions of the country and, therefore, with very different characteristics, both in terms of size and technological, technical and operational capacity; therefore, forms are also sent on printed paper and on magnetic media.

To make it even easier to fill in, service providers, such as those that only supply water and do not provide sewage services, only receive questionnaires about the area in which they operate, and the same applies to those municipalities that do not have a service concession and do not need to fill in information about this.

The information requested is the most varied, from the constitution of the service provider, its legal characteristics, city, scope, etc., to registration, operational, financial and quality indicators, all grouped by subject.

And in each set, there is a huge range of indicators that must be filled in by the service providers, after which the data is sent to the SNIS, and there is a

checking the consistency of the data, which is tabulated and submitted to the service providers for analysis, criticism and suggestions, for subsequent publication.

By way of illustration, in the 2002 SNIS, 279 service providers made up the sample with the following composition: 25 regional providers, 6 micro-regional providers and 248 local providers. Together, these providers are responsible for the water supply services of 4,186 Brazilian municipalities, corresponding to 94.3% of the national urban population, making this a representative sample.

3.3 Questioning SNIS

Despite this, there is a question mark that must be raised regarding the reliability of the SNIS information, despite all the technical work carried out by its developers and planners, with investments in technology to obtain the data, with modern computer programs and tabulation techniques. The sample is not 100% reliable, so we must analyze the data carefully.

Under no circumstances is it being said that the SNIS does not portray the Brazilian reality of the water and sewage sector, nor that it is not valid or useful, quite the contrary, what is being raised is the fact that, as the information is obtained through responses from service providers, in other words, the entity receives the questionnaire and has to answer it, then send it to SNIS, without visits or even inspections of the veracity of the information, it may happen that some indicators are divergent from reality.

This can happen for various reasons, including the wrong way of obtaining the data, technical and operational incapacity to measure the indicators, the service provider's lack of commitment to the information, the service provider's interest in manipulating the information for their own benefit and the lack of data.

A service provider may not use a very correct way of obtaining the information, in other words, use a technique that does not accurately measure the data they provide, causing divergences between the true value and the one provided, which may not be due to bad faith, but due to problems in obtaining the indicators.

Technical and operational incapacity can be considered as factors that harm the indicators, since the smaller water and sewage service providers, with few financial and technical resources, suffer from a lack of qualified staff and their own equipment. For example, in order to calculate losses, the volume of water produced must be measured by macro-meters, and that consumed by residents by micro-meters.

In addition to the equipment, it is necessary to have personnel to measure and calculate the data, as well as, of course, a technical study by qualified professionals, in order to size the best meter, of which there are several, each one for an average volume and also for the size of the network, since a wrongly sized meter gives values that differ from the real one.

Therefore, small service providers who don't have the resources to invest in these areas, and who don't have technically qualified staff, may calculate incorrect values and report them in SNIS as the real ones, damaging the sample.

The service provider may also have no direct benefit in answering the questionnaire, which is too complex and too large, and may report average values, or even values that are out of date from some other calculation in the past, in order to get out of the commitment to answer the questionnaire, which also leads to problems with the reliability of the final result. Although this may seem absurd, it really does happen, because it's worth remembering that there is more to SNIS than the service provider answering the questionnaire.

The administrator of a water and sewage system, in addition to his day-to-day administrative problems, such as purchasing, human resources, the warehouse, dealing with the public, complying with the Fiscal Responsibility Law, political pressure from the mayor, etc., also suffers from the bureaucratic commitments of the public administration, and various other types of questioning, such as from the SEADE Foundation, IDEC, the Court of Auditors and others.

There is also the fact that some service providers try to alter their information in the naive sense that the data could harm or benefit them. One example is that, in an analysis of the release of funds for investments in two municipalities, the analysis is carried out using the indicators submitted to SNIS, and so a good or bad indicator can favor or harm a particular service provider in the election in question.

Finally, there are the service providers who do not have concrete or even estimated data for their SNIS information, leaving gaps in the questionnaire. However, this is a range of very valuable information and is the basis of the work.

3.4 Selected Indicators

SNIS presents various indicators in the following groups:

- Economic, financial and administrative indicators;
- Operational Indicators - Water and Sewage;
- Quality indicators.

Each of the four groups has several indicators; for this study, the following indicators were selected from the groups:

Table 3.1

Economic-Financial and Administrative Indicators

Item	Definition	Express
1	Productivity Index: Active Savings per Own Personnel	Economy/employee
	Number of Active Savings (Water + Sewage) Total Number of Own Employees	
2	Average Fare	R$/$m^3$
	Total Expenditure on Services Total Volume Billed (Water + Sewage)	
3	Impact of Personnel Expenses and Third-Party Services on Total Expenses for Services	percentage
	Own Personnel Expenses + External Services Expenses	
4	Total Expenditure on Services Average Annual Expenditure per Employee	R$/employee
	Staff costs	
5	Total Number of Own Employees Productivity Index: Active Savings per Total Staff (Equivalent)	Savings/Employment Equi.
	Total Number of Active Savings (Water+Sewage)	
6	Equivalent Number of Total Personnel Productivity Index: Own Employees per thousand connections (water + sewage)	Employees/thousand lig.
	Total Number of Own Employees Number of Active Connections (Water+Sewage)	

Source: SNIS 2001.

Table 3.2

Operational Indicators - Water and Sewage

Item	Definition	Express
1	Hydrometric index	Percentage
	Number of Micrometered Water Connections	

	Quantity of Active Water Connections	
2	Macrometering index	Percentage
	Macrometered Water Volume	
	Volume of Water Available for Distribution	
3	Billing Loss Index	Percentage
	Vol. of Water (Produced+Treated - Service) - Vol. of Water Billed	
	Vol. Of Water (Produced+Treated Imported - of service_	
4	Sewage Collection Index	Percentage
	Volume of Sewage Collected	
	Volume of Water Consumed-Volume of Treated Water Exported Sewage	
5	Treatment Index	Percentage
	Volume of Sewage Treated	
6	Volume of Sewage Collected Total Water Service Index	percentage
	Pop. Served with Water Supply Water Supply	
	Pop. Total Pop. Served by Water Supply Water Supply	

Source: SNIS 2001.

Table 3.3

Quality indicators

Item	Definition	Express
1	Average duration of stoppages	Hours/work stoppages
	Duration of stoppages Number of stoppages	
2	Economies hit by intermittency	Savings/interruption
	Quant. Of Active Savings Affected by Intermittency	
	Prolonged	
	Number of Systematic Interruptions	
3	Incidence of Fecal Coliform Analysis Out of Standard	Percentage
	Quant. of Samples for Fecal Coliform Analysis with	
	Out-of-Standard Results Number of Samples Analyzed for Fecal Coliforms	

4	Average duration of services performed Service Execution Time Number of Services Performed	Time/service

The study in question will focus on and analyze the above indicators for municipalities in the Southeast region, separating public and private service providers.

The municipalities, as well as the bodies studied, are shown in the table below, remembering that they are all in the Southeast region.

Table 3.4

Municipalities and Service Providers

Qty.	Municipalities	Name of service provider
1	Americana-SP	Water and Sewerage Department
2	Araçatuba-SP	Araçatuba Water and Sewage Department
3	Aracruz-ES	Autonomous Water and Sewage Service
4	Araguari-MG	Water and Sewage Superintendence
5	Araraquara-SP	Autonomous Department of Water and Sewage
6	Barra Mansa-RJ	Autonomous Water and Sewage Service of Barra Mansa
7	Barretos-SP	Autonomous Water and Sewage Service of Barretos
8	Bauru-SP	
9	Birigui-SP	Water and Sewerage Department Birigui Water and Sewerage Department
10	Cachoeiro de Itapemirim-ES	Aguas de Cocheiro SA
11	Caeté-MG	Autonomous Water and Sewage Service
12	Campos dos Goytacazes-RJ	Aguas do Paraiba AS
13	Catanduva-SP	Municipal Sanitation Department
14	Coqueiral-MG	Autonomous Water and Sewage Service
15	Cosmópolis-SP	Water and Sewerage Department
16	Dracena-SP	Dracena Water, Sewage and Pavement Development Company (EMDAEP)
17	Engenheiro Coelho-SP	Water and Sewerage Department
18	Governador Valadares-MG	Autonomous Water and Sewage Service
19	Guarâ-SP	Aguas de Guarà Ltda
20	Guaratinguetâ-SP	Guaratinguetà Autonomous Water and Sewage Service
21	Guarulhos-SP	Autonomous Water and Sewage Service
22	Holambra-SP	Holambra City Hall
23	Indaiatuba-SP	Autonomous Water and Sewage Service
24	Itabira-MG	Autonomous Water and Sewage Service
25	Itaguara-MG	Autonomous Water and Sewage Service
26	Itaûna-MG	Autonomous Water and Sewage Service
27	Itu-SP	Autonomous Water and Sewage Service of Itu
28	Ituiutaba-MG	Water and Sewage Superintendence of Ituiutaba
29	Jacarei-SP	Autonomous Water and Sewage Service of Jacarei
30	Jaguariûna-SP	Municipal Sanitation Department
31	Jaù-SP	Water and Sewerage Service of the Municipality of Jaù
32	Jerônimo Monteiro-ES	Autonomous Water and Sewage Service
33	Leme-SP	Superintendence of Water and Sewage of the City of Leme
34	Linhares-ES	Linhares Autonomous Water and Sewage Service
35	Limeira-SP	Aguas de Limeira S/A
36	Mairinque-SP	Concessionària de Aguas de Mairinque Ltda
37	Marilia-SP	Marilia Water and Sewage Department
38	Mauà-SP	Basic Sanitation in the Municipality of Mauà
39	Mogi Guaçu-SP	Autonomous Municipal Water and Sewage Service of Mogi Guaçu
40	Moji das Cruzes-SP	Municipal Water and Sewage Service
41	Moji-Mirim-SP	Autonomous Water and Sewage Service
42	Muraé-MG	Municipal Department of Urban Sanitation
43	Niterói-RJ	Aguas de Niterói AS
44	Nova Friburgo-RJ	Concessionària de Aguas e Esgotos de Nova Friburgo Ltda
45	Ourinhos-SP	Superintendence of Water and Sewage of Ourinhos
46	Ouro Verde-SP	City Hall
47	Passos-MG	Autonomous Water and Sewage Service
48	Paulicéia-SP	City Hall

49	Quarry-SP	Urban Services Department
50	Petrópolis-RJ	Aguas do Imperador AS
51	Piracicaba-SP	Municipal Water and Sewage Service
52	Pirassununga-SP	Pirassununga Water and Sewage Service
53	Poços de Caldas-MG	Municipal Water and Sewage Department
54	Ponte Nova-MG	Municipal Department of Water and Sewerage and Sanitation
55	Ribeirao Preto-SP	Ribeirao Preto Water and Sewage Department
56	Rio Claro-SP	Autonomous Water and Sewage Department
57	Sacramento-MG	Autonomous Water and Sewage Service
58	Salto-SP	Water and Sewerage Service
59	Santa Bàrbara d'Oeste-SP	Water and Sewerage Department
60	Santo André-SP	Santo André Municipal Environmental Sanitation Service
61	Sao Bernardo do Campo-SP	Water and Sewerage Department
62	Sao Caetano do Sul-SP	Sao Caetano do Sul Water and Sewage Department
63	Sao Carlos-SP	Autonomous Water and Sewage Service
64	Sao Joao do Pau d'Alho-SP	City Hall
65	Sao José do Rio Preto-SP	Autonomous Water and Sewage Service
66	Sao Mateus-ES	Autonomous Water and Sewage Service
67	Sete Lagoas-MG	Autonomous Water, Sewage and Urban Sanitation Service
68	Sorocaba-SP	Autonomous Water and Sewage Service
69	Sumaré-SP	Municipal Water and Sewage Department
70	Tupi Paulista-SP	City Hall
71	Uberaba-MG	Uberaba Development and Sanitation Operational Center
72	Uberlândia-MG	Municipal Water and Sewage Department
73	Unai-MG	Autonomous Water and Sewage Service
74	Valinhos-SP	Valinhos Water and Sewage Department
75	Vinhedo-SP	Water and Sewage Department
76	Volta Redonda-RJ	Autonomous Water and Sewage Service of Volta Redonda

Source: SNIS 2001, prepared by the author.

All the municipalities listed in the table are those that have their information on basic sanitation published in the 2001 SNIS.

3.5 Relationship between indicators - Public vs. Private

We will present the indicators for each of the municipalities studied in the 2001 SNIS in the south-eastern region of Brazil; firstly, the municipalities will be grouped into those with public services and those with private services. Mixed company service providers with public administration were not considered in this study.

It should be noted that service providers who operate in more than one location were also disregarded in this study, in order to give greater equality between municipalities when comparing indicators.

The services of a public nature include all those that are administered directly or by an autarchy. The table below shows the municipalities and bodies that provide public basic sanitation services.

Table 3.5

Public Service Providers

Qty.	City	Name	Nature
1	Americana	Water and Sewerage Department	Autarchy
2	Araçatuba	Araçatuba Water and Sewage Department	Autarchy
3	Aracruz	Autonomous Water and Sewage Service	Autarchy
4	Araguari	Water and Sewage Superintendence	Autarchy
5	Araraquara	Autonomous Department of Water and Sewage	Autarchy
6	Barra Mansa	Autonomous Water and Sewage Service of Barra Mansa	Autarchy
7	Barretos	Autonomous Water and Sewage Service of Barretos	Autarchy
8	Bauru	Water and Sewerage Department	Autarchy

9	Birigui	Birigui Water and Sewage Department	Public Administration Direct
10	Caeté	Autonomous Water and Sewage Service	Autarchy
11	Catanduva	Municipal Sanitation Department	Public Administration Direct
12	Coconut grove	Autonomous Water and Sewage Service	Autarchy
13	Cosmopolis	Water and Sewerage Department	Public Administration Direct
14	Dracena	Empresa de Des., Agua, Esg. e Pav. Dracena (EMDAEP)	Autarchy
15	Engenheiro Coelho	Water and Sewerage Department	Public Administration Direct
16	Governador Valadares	Autonomous Water and Sewage Service	Autarchy
17	Guaratinguetà	Autonomous Water and Sewage Service of Guaratinguetà	Autarchy
18	Guarulhos	Autonomous Water and Sewage Service	Autarchy
19	Holambra	Holambra City Hall	Public Administration Direct
20	Indaiatuba	Autonomous Water and Sewage Service	Autarchy
21	Itabira	Autonomous Water and Sewage Service	Autarchy
22	Itaguara	Autonomous Water and Sewage Service	Autarchy
23	Itaûna	Autonomous Water and Sewage Service	Autarchy
24	Itu	Autonomous Water and Sewage Service of Itu	Autarchy
25	Ituiutaba	Water and Sewage Superintendence of Ituiutaba	Autarchy
26	Jacarei	Autonomous Water and Sewage Service of Jacarei	Autarchy
27	Jaguariûna	Municipal Sanitation Department	Public Administration Direct
28	Jaù	Water and Sewerage Service of the Municipality of Jaù	Autarchy
29	Jerônimo Monteiro	Autonomous Water and Sewage Service	Autarchy
30	Rudder	Superintendência de Agua e Esgotos da Cidade de Leme	Autarchy
31	Linhares	Linhares Autonomous Water and Sewage Service	Autarchy
32	Marilia	Marilia Water and Sewage Department	Autarchy
33	Mauà	Basic Sanitation in the Municipality of Mauà	Autarchy
34	Mogi Guaçu	Mogi Municipal Water and Sewage Service Guaçu	Autarchy
35	Moji das Cruzes	Municipal Water and Sewage Service	Autarchy
36	Moji-Mirim	Autonomous Water and Sewage Service	Autarchy
37	Muriaé	Municipal Department of Urban Sanitation	Autarchy
38	Ourinhos	Superintendence of Water and Sewage of Ourinhos	Autarchy
		Public Administration	
39	Green Gold	City Hall	Direct
40	Steps	Autonomous Water and Sewage Service	Autarchy
41	Paulicéia	City Hall	Public Administration Direct
42	Quarry	Urban Services Department	Public Administration Direct
43	Piracicaba	Municipal Water and Sewage Service	Autarchy
44	Pirassununga	Pirassununga Water and Sewage Service	Autarchy
45	Poços de Caldas	Municipal Water and Sewage Department	Autarchy
46	Ponte Nova	Department of Water, Sewerage and Sanitation	Autarchy
47	Ribeirao Preto	Ribeirao Preto Water and Sewage Department	Autarchy
48	Rio Claro	Autonomous Water and Sewage Department	Autarchy
49	Sacramento	Autonomous Water and Sewage Service	Autarchy
50	Jump	Water and Sewerage Service	Public Administration Direct
51	Santa Bàrbara d Oeste	Water and Sewerage Department	Autarchy
52	Saint André	Santo Municipal Environmental Sanitation Service André	Autarchy
			Public

53	Sao Bernardo do Campo	Water and Sewage Department	Administration Direct
54	Sao Caetano do Sul	Sao Caetano do Sul Water and Sewage Department	Autarchy
55	Sao Carlos	Autonomous Water and Sewage Service	Autarchy
56	Sao Joao do Pau d Alho	City Hall	Public Administration Direct
57	Sao José do Rio Preto	Autonomous Water and Sewage Service	Public Administration Direct
58	Sao Mateus	Autonomous Water and Sewage Service	Autarchy
59	Sete Lagoas	Autonomous Water, Sewage and Sanitation Service Urban	Autarchy
60	Sorocaba	Autonomous Water and Sewage Service	Autarchy
61	Sumaré	Municipal Water and Sewage Department	Autarchy
62	Tupi Paulista	City Hall	Public Administration Direct
63	Uberaba	Development and Sanitation Operational Center of Uberaba	Autarchy
64	Uberlândia	Municipal Water and Sewage Department	Autarchy
65	Unai	Autonomous Water and Sewage Service	Autarchy
66	Valinhos	Valinhos Water and Sewage Department	Autarchy
67	Vineyard	Water and Sewage Department	Public Administration Direct
68	Volta Redonda	Autonomous Water and Sewage Service of Volta Redonda	Autarchy

Source: SNIS 2001, prepared by the author.

The table below lists the service providers, as well as the municipalities that have private companies providing services.

Table 3.6

Private Service Providers

Qty.	City	Name	Nature
1	Cachoeira de Itapemirim	Aguas de Cachoeiro S/AS	Private Company
2	Campos dos Goytacazes	Aguas do Paraiba S/A	Private Company
3	Guarà	Aguas de Guarà Ltda	Private Company
4	Limeira	Aguas de Limeira S/A	Private Company
5	Mairinque	Ciàgua Concessionària de Aguas de Mairinque Ltda	Private Company
6	Niteóri	Aguas de Niterói S/A	Private Company
7	Nova Friburgo	Nova Friburgo Water and Sewage Concessionaire Ltda	Private Company
8	Petrópolis	Aguas do Imperador S/A	Private Company

Source: SNIS 2001, prepared by the author.

In this way, all the indicators selected for each of the municipalities listed were calculated, separated into two groups, as above, i.e. public and private, and at the end of the table there is a totalizer of the indicators, which are calculated as the average values of the group, and not as the average of the values of the group (the method applied by SNIS).

With this, in all the components of each indicator of a group, in this case public and private, the values corresponding to each of the service providers are added up and the indicator is calculated.

As a result of this calculation method, the totals at the end of the tables only take into account the service providers who have submitted all the information required for the calculation, i.e. if, when calculating a indicator, a service provider submits data that is not available (blank field), this is disregarded when calculating the totals for the indicator in question.

To begin the comparisons between the municipalities, firstly, the Economic-Financial and Administrative indicators are presented, as presented above, following the same identification numbering in the table (columns). Table 3.7 shows the indicators for municipalities with public service providers and Table 3.8 shows the indicators for private service providers.

Table 3.7

Economic-Financial and Administrative Indicators

Public Service Providers

Qty.	City	1	2	3	4	5	6
1	Americana	410	0,71	63,00	13.600,48	244	3,2
2	Araçatuba	345	0,71	64,5	28.248,72	194	3,3
3	Aracruz	239	0,43		8.942,89	146	4,6
4	Araguari	509		58,9	9.036,67	264	2,0
5	Araraquara	558	0,57	59,9	29.409,56	443	2,0
6	Barra Mansa	267	0,56	76,8	13.545,02	176	6,0
7	Barretos	285	0,63	72,2	15.604,16	256	3,7
8	Bauru	339	0,73	67,7	16.315,77	269	3,5
9	Birigui	552	0,23	68,3	7.914,39	204	2,0
10	Caeté	206	0,53	63,3	12.119,97	156	5,2
11	Catanduva	822	0,35	44,0	12.633,04	519	1,3
12	Coconut grove	220	0,34	82,7	9.399,91	153	5,0
13	Cosmopolis	362	0,35		9.383,72	289	2,8
14	Dracena		0,60				
15	Engenheiro Coelho				8.835,33		2,5
16	Governador Valadares	322	0,48		9.210,63	192	4,0
17	Guaratinguetà	387	0,57	79,4	11.069,70	216	3,0
18	Guarulhos	334		53,8	32.594,02	273	3,8
19	Holambra		0,63		6.433,82		3,7
20	Indaiatuba	330	0,56	63,2	16.254,65	227	3,2
21	Itabira	169	0,43	66,0	10.719,89	134	7,3
22	Itaguara	209	0,33	84,5	11.588,56	177	5,2
23	Itaûna	298	0,52	39,5	7.738,03	209	4,5
24	Itu	240	0,94	71,11	10.488,04	139	4,5
25	Ituiutaba	368	0,49	81,2	14.263,73	203	3,1
26	Jacarei	338	0,84	46,2	15.881,93	320	3,4
27	Jaguariûna		0,92		13.985,54		3,0
28	Jaù	627	0,53	37,4	14.058,47	355	1,6
29	Jerônimo Monteiro	424		80,6	19.815,10	349	2,8
30	Rudder	406	0,56	33,6	9.793,10	137	2,5
31	Linhares	296	0,49	72,1	10.139,18	202	4,3
32	Marilia	330	0,77	54,9	14.948,04	220	3,5
33	Mauà	1.025	1,08	26,4	20.864,71	661	1,3
34	Mogi Guaçu	655	0,69	48,4	15.092,76	634	1,9
35	Moji das Cruzes	921	0,98	39,4	29.334,24	379	1,2
36	Moji-Mirim	376	0,64	48,2	17.226,10	278	3,1
37	Muriaé		0,68	56,9			
38	Ourinhos	228	0,46	42,9	7.691,81	192	5,0
39	Green Gold		0,23		1.407,15		3,7
40	Steps	640	0,26	65,7	19.987,02	481	1,8
41	Paulicéia						
42	Quarry	456		39,8	8.513,01	430	2,3
43	Piracicaba	447	0,65	74,5	22.028,79	254	2,6
44	Pirassununga	254		70,0	14.099,54	203	4,1
45	Poços de Caldas	301	0,76	59,8	19.298,04	274	4,1
46	Ponte Nova	255		65,9	13.018,09	209	5,6
47	Ribeirao Preto		0,80		17.260,86		3,4
48	Rio Claro	420	0,65	74,0	22.089,49	297	2,6

49	Sacramento	336		58,9	12.978,21	226	3,1
50	Jump		0,38	43,3			
51	Santa Bàrbara d Oeste	562	0,75	73,9	21.852,64	270	2,1
52	Saint André	350	0,87	55,5	15.460,60	142	4,5
53	Sao Bernardo do Campo	1.03	1,19	28,5	19.650,41	705	1,4
54	Sao Caetano do Sul	573	0,98				3,2
55	Sao Carlos	344	0,65	40,7	17.415,55	291	3,1
56	Sao Joao do Pau d Alho				4.785,34		4,5
57	Sao José do Rio Preto		0,21	30,0	563,61		
58	Sao Mateus	256		81,0	11.992,50	175	4,3
59	Sete Lagoas	360	0,51	62,1	9.833,38	170	3,3
60	Sorocaba	332	0,59	58,3	21.068,99	250	3,4
61	Sumaré	351			22.061,75	258	2,9
62	Tupi Paulista				9.133,06		2,1
63	Uberaba	377	0,44	58,4	12.063,91	210	3,2
64	Uberlândia	585	0,29	76,5	11.861,38	320	2,6
65	Unai	585	0,41	71,5	18.154,26	347	2,0
66	Valinhos	294	0,81	72,8	18.268,10	225	4,3
67	Vineyard	294	1,14	43,9	17.166,54	188	4,1
68	Volta Redonda	388	0,44	56,0	19.748,28	316	4,0
	Total	400	0,65	55,3	17.486,67	231	3,1

Source: SNIS 2001, prepared by the author.

The economic, financial and administrative indicators for private service providers are presented below, following the same parameters as for public services.

Table 3.8

Economic-Financial and Administrative Indicators

Private Service Providers

Qty.	City	1	2	3	4	5	6
1	Cachoeira de Itapemirim	478	0,88	35,2	21.137,18	456	3,1
2	Campos dos Goytacazes	479	1,00	32,7	15.133,90	370	3,3
3	Guarà	492	0,41	29,3	8.533,33	404	2,2
4	Limeira	888	0,58	19,6	20.882,22	748	1,3
5	Mairinque	318	0,88	28,1	15.633,11	259	3,8
6	Niteóri	768	1,10	29,2	22.584,07	345	3,7
7	Nova Friburgo	443	0,41	41,9	13.832,03	390	3,5
8	Petrópolis	612	0,87	31,0	19.370,05	398	2,8
	Total	600	0,83	29,5	18.568,22	399	2,8

Source: SNIS 2001, prepared by the author.

For an easier visualization of the comparison between the types of service providers, Table 3.9 shows the description of each of the six indicators studied and the average value for each group, public and private.

Table 3.9

Economic-Financial and Administrative Indicators

Comparison: Public vs. Private

INDICATORS	PUBLIC	PRIVATE
1 - Active savings by own staff (Economy/Employee)	400	600
2 - Average Fare (R$/m^3)	0,65	0,83
3 - Impact of Personnel Expenses and Third-Party Services on Total Expenses for Services (percentage)	55,3%	29,5%
4 - Average Annual Expenditure per Employee	17.486,67	18.568,22

(R$/Employee)		
5 - Active Savings by Total Personnel (Economy/Employment)	231	399
6 - Own employees per thousand connections (Employees/thousand connections)	3,1	2,8

Source: SNIS 2001, prepared by the author.

Following the comparative table, some statements and comments can be made about each of the indicators. When analyzing the first, i.e. Active Savings per Own Staff, it can be seen that the private sector has 50% more savings than the public sector (600 to 400), demonstrating greater efficiency, i.e. the private sector has more clients than the public sector, when analyzing the number of employees.

The term savings is understood to be not exactly the number of connections with water meters, but the total water connections; to clarify further, imagine a building with 20 apartments, which has only one water connection, and thus only one water meter; however, when analyzing consumption, a total of twenty savings is considered.

A similar result can be seen when analyzing indicator 5 - Active Savings, i.e. by analyzing only the savings that are active, the result is almost 73% better for private services, again demonstrating greater efficiency.

It can also be seen that private service providers have fewer of their own employees per thousand calls, perhaps due to the legislation governing public services, the characteristic of having employees under the statutory system, with stability, which ends up causing a swelling of the public machinery, or even the use of some municipalities as "job hangers" by mayors and public administrators.

One question that could be raised is that, instead of using their own employees, private service providers use outsourced labor, which would certainly not be captured by indicator 6, but by indicator 3 - Incidence of Personnel Expenditure and Outsourced Services in Total Expenditure on Services. This hypothesis doesn't hold up, because private companies spend 29.5% of their total expenditure on services on their own or outsourced staff, while public companies spend 55.3%, showing a certain inefficiency on the part of public companies, because according to indicator 4 - Average Annual Expenditure per Employee - we see that private companies spend more than public companies on average over the year, showing that they invest more in their own staff than public companies.

In this first stage, the indicators are in favor of private service providers, showing greater efficiency, but one indicator, 2 - Average Tariff Practiced - is unfavorable to the private sector, as it indicates that they have a higher tariff than the public sector, of approximately 27%.

Thus, after analyzing the economic, financial and administrative indicators of public and private service providers in the Southeast, it can be said that the private ones are more efficient because they have better indicators, but this efficiency is marked by higher tariffs for their services.

After analyzing the first group of indicators, Tables 3.10 and 3.11 will show the operational indicators for both water and sewage, presented in the same way as above, i.e. one table for public service providers and another for private ones.

Table 3.10

Operational Indicators

Public Service Providers

Qty.	City	1	2	3	4	5	6
1	Americana	100,0	100,0	50,6	87,1	90,7	99,6
2	Araçatuba	100,0		55,8		100,0	99,3
3	Aracruz	98,4				59,7	100,0
4	Araguari	24,0					95,9
5	Araraquara	98,8	100,0	38,1	80,0	161,0	100,0
6	Barra Mansa	90,0	100,0	27,7	80,0	0	94,8
7	Barretos	100,0	0	27,8	80,0	50,0	94,2
8	Bauru	99,2	100,0	44,7	80,0	0	98,3
9	Birigui	98,9	17,7	50,2	88,8	0	,100,0
10	Caeté	87,3				0	100,0
11	Catanduva	100,0	0	47,6	85,0	4,6	98,2
12	Coconut grove	98,7	0	33,0	80,0	50,0	83,6
13	Cosmopolis	18,1	0	45,2	76,8	10,3	96,4
14	Dracena	95,6				20,3	91,6
15	Engenheiro Coelho	100,0					74,6
16	Governador Valadares	85,1					95,8
17	Guaratinguetà	96,0	0	32,3	96,0	10,0	98,5
18	Guarulhos	98,6	100,0	53,2			91,4
19	Holambra	100,0				100,0	53,1
20	Indaiatuba	99,1	97,7	29,7	99,4	7,7	91,5
21	Itabira	100,0	73,0	51,0	72,6	7,9	91,2
22	Itaguara	94,6	0	10,2	80,4	0	80,7
23	Itaûna	99,7	100,0	35,1	80,1	0	97,3
24	Itu	100,0			99,1	100,0	100,0
25	Ituiutaba	100,0	100,0	23,1	76,7	64,9	94,2
26	Jacarei	99,1			80,0	1,9	100,0
27	Jaguariûna	99,3	63,8	76,1	80,0	0	100,0
28	Jaù	100,0	0	53,8	76,3	0	100,0
29	Jerônimo Monteiro	100,0					82,8
30	Rudder	96,0	0	26,6	82,5	0	100,0
31	Linhares	99,3	100,0	17,8	58,3	20,4	82,3
32	Marilia	92,1	9,7	47,5	80,0	1,9	96,6
33	Mauâ	100,0			68,3	0	96,6
34	Mogi Guaçu	92,2	100,0	49,3	81,4	56,7	100,0
35	Moji das Cruzes	88,1	0	60,5	99,1	0	89,6
36	Moji-Mirim	100,0				1,9	87,5
37	Muriaé	89,5	100,0	23,6	11,8	3,6	100,0
38	Ourinhos	96,8	100,0	42,5	95,1	86,6	95,2
39	Green Gold	0,4				100,0	100,0
40	Steps	100,0	0	12,1	80,0	0	93,3
41	Paulicéia	74,2	100,0	0			
42	Quarry	100,0	0	12,6			99,5
43	Piracicaba	100,0	100,0	38,4	99,3	29,7	100,0
44	Pirassununga	100,0		40,5	80,0		93,1
45	Poços de Caldas	99,8	98,3	42,1	83,2	17,0	97,6
46	Ponte Nova	85,9	100,0			0	98,8
47	Ribeirao Preto		100,0	61,6	69,6	16,0	97,6
48	Rio Claro	95,2	100,0	37,0	80,0	11,1	100,0
49	Sacramento	98,6	4,3	40,7			74,1
50	Jump	100,0	100,0	30,1	91,8	0	100,0
51	Santa Bàrbara d Oeste	100,0	100,0	29,8	79,4	0	100,0
52	Saint André	96,7	100,0	28,7	61,8	0	97,6
53	Sao Bernardo do Campo	94,4	100,0	49,5	84,5	0,3	99,9
54	Sao Caetano do Sul	100,0	100,0	27,8	102,6	25,0	100,0
55	Sao Carlos	98,8	94,6	47,0	129,3	3,0	93,2
56	Sao Joao do Pau d Alho	15,1	0				100,0
57	Sao José do Rio Preto	99,2	84,5	50,1	95,2	4,1	100,0
58	Sao Mateus	93,6					74,7

59	Sete Lagoas	95,8	0	61,8	88,3	10,7	100,0
60	Sorocaba	100,0	96,5	33,3	77,1	1,7	99,9
61	Sumaré	99,5	84,7	55,2	79,5	2,0	96,7
62	Tupi Paulista	100,0					78,1
63	Uberaba	99,4	97,6	31,6	157,6		98,1
64	Uberlândia	100,0	100,0	37,7	80,0	30,7	96,3
65	Unai	97,9	0	18,5	68,0	100,0	80,7
66	Valinhos	99,8			80,0	0	90,5
67	Vineyard	100,0			47,2	0	100,0
68	Volta Redonda	98,4	100,0	50,0	80,0	13,7	100,0
	Total	96,3	82,1	45,1	84,0	19,3	98,0

Source: SNIS 2001, prepared by the author.

Below are the water and sewage operational indicators for private service providers, following the same parameters as the public ones.

Table 3.11

Operational Indicators

Private Service Providers

Qty.	City	1	2	3	4	5	6
1	Cachoeira de Itapemirim	92,5	81,3	29,2	80,0	7,0	95,9
2	Campos dos Goytacazes	71,3	0	48,1	42,8	0	76,8
3	Guarà	50,4	100,0	20,0	87,4	4,1	93,7
4	Limeira	100,0	96,0	12,9	80,0	80,5	94,0
5	Mairinque	99,2		37,7	77,0	0	81,4
6	Niteóri	78,2	100,0	27,4	74,5	90,2	94,8
7	Nova Friburgo	63,6	100,0	11,0	73,6		88,9
8	Petrópolis	94,9	100,0	7,3	57,3	25,6	72,8
	Total	83,8	78,2	26,3	69,8	60,1	66,4

Source: SNIS 2001, prepared by the author.

By regrouping the average values obtained in the two tables above for the six indicators studied so far, the following is obtained:

Table 3.12

Operational Indicators

Comparison: Public vs. Private

INDICATORS	PUBLIC	PRIVATE
1 - Hydrometric index (Percentage)	96,3	83,8
2 - Macrometering index (Percentage)	82,1	78,2
3 - Billing Loss Index (percentage)	45,1	26,3
4 - Sewage Collection Index (Percentage)	84,0	69,8
5 - Sewage Treatment Index (Percentage)	19,3	60,1
6 - Total Water Service Index (Percentage)	98,0	66,4

Source: SNIS 2001, prepared by the author.

When we evaluate these indicators, at first glance, the end result is that public service providers are more efficient than private ones, because, in a way, they are more efficient at providing services than their private competitors.

This is because, in the first indicator, 1 - Hydrometric Index, which describes the current situation of the system, i.e. how many percent are covered with hydrometers, we can see that the index is extremely favorable to public services, which reach close to 100%, and the private ones far behind.

As for the second indicator, 2 - the Macrometering Index, which measures whether or not all the water produced is being metered, the indices for both service providers are close, with no major differences.

Indicator 3 - Losses, which indicates the efficiency or otherwise of the system, shows that the private sector's index is 20 percentage points better than the public sector's, but this may reflect the fact that the private sector has a lower micro-metering than the public sector, as shown in indicator 1.

Indicators 4 and 5 are somewhat curious, because in indicator 4 - Sewage Collection Index - the public sector has an advantage over the private sector of just over 15 percentage points, i.e. the public sector collects more sewage than the private sector, but in indicator 5 - Sewage Treatment Index - the private sector has an advantage of over 40 percentage points, showing that although they collect less sewage, they treat more of it.

Finally, in indicator 6 - Total Water Service Index - the public sector has the upper hand, with an index of almost 100%. At the end of the analysis, it can be assumed that the publics are really more efficient, but, as already discussed, the data was obtained from the SNIS, and it may have some deficiencies that may be present here, i.e. the data may not reflect reality.

This is because all the data provided is the responsibility of the service provider, and so there is a good chance that this data does not reflect reality, i.e. it may have been used for political purposes, or even camouflaged to avoid criticism, especially from public service providers. We can't say that this has happened, but this is a caveat and a warning for when SNIS data is analyzed.

Another hypothesis for this curiosity in the indicators presented here lies in the fact that there may have been some kind of concession process in order for these imperfections to occur, i.e. privatizing a service that is being carried out inefficiently by the state, because it doesn't have the financial resources to invest and thus make it possible to improve services. Therefore, privatization may have taken place, but it is still in the early stages of analyzing its results.

However, considering that the data is indeed true and reliable, it can be concluded that, in terms of operability, public service providers have a slight advantage over private ones, with the caveat mentioned above.

The last group of indicators, the quality indicators, will now be analyzed. Four indicators will be presented in the same way as the previous indicators, following the division between public and private.

Table 3.13

Quality indicators

Public Service Providers

Qty.	City	1	2	3	4
1	Americana	13,00			
2	Araçatuba			0,0014	
3	Aracruz				
4	Araguari				
5	Araraquara			0,0432	
6	Barra Mansa	10,00		0,0439	

7	Barretos				
8	Bauru				
9	Birigui	42,25			
10	Caeté	8,00	16,67	0,0208	1,05
11	Catanduva				1,49
12	Coconut grove				
13	Cosmopolis	10,00			
14	Dracena				
15	Engenheiro Coelho	2,00		0,0096	
16	Governador Valadares				
17	Guaratinguetà	10,17	1.906,69	0,8232	1,43
18	Guarulhos			0,0012	
19	Holambra				
20	Indaiatuba				
21	Itabira				
22	Itaguara	13,20		0,0146	7,82
23	Itaûna	9,71	635,33	0,2492	0,50
24	Itu	0,0111			
25	Ituiutaba	31,00			
26	Jacarei	0,0370	2,45		
27	Jaguariûna	8,00	0,25		
28	Jaù	0,1000			
29	Jerônimo Monteiro				
30	Rudder	1,10	0,88		
31	Linhares	7,50			
32	Marilia	27,25	277,89		5,00
33	Mauà	7,00			
34	Mogi Guaçu				
35	Moji das Cruzes				
36	Moji-Mirim				
37	Muriaé	106,67			0,50
38	Ourinhos				
39	Green Gold	4,00			
40	Steps	6,00	0,50		
41	Paulicéia	7,75			
42	Quarry				
43	Piracicaba	6,33	0,04		
44	Pirassununga	3,33	1,0000		
45	Poços de Caldas	6,00	0,0058	0,50	
46	Ponte Nova	5,00	158,02		
47	Ribeirao Preto	0,0030			
48	Rio Claro				
49	Sacramento				
50	Jump	0,0083	1,50		
51	Santa Bàrbara d Oeste	13,33			
52	Saint André	9,75			
53	Sao Bernardo do Campo	14,13	6.624,47	0,0168	
54	Sao Caetano do Sul	3,89			
55	Sao Carlos	0,0040			
56	Sao Joao do Pau d Alho				
57	Sao José do Rio Preto	18,27	99,21		
58	Sao Mateus	5,00	0,0370		
59	Sete Lagoas	3,42	3,60	0,0254	1,30
60	Sorocaba	0,0007			
61	Sumaré	18,75			
62	Tupi Paulista				
63	Uberaba	6,00	100,00	0,0667	
64	Uberlândia				
65	Unai	19,50	3.833,33	0,0262	0,30
66	Valinhos	10,00	1,90		
67	Vineyard	0,00			
68	Volta Redonda	26,00			
	Total	8,54	96,73	0,0590	0,91

Source: SNIS 2001, prepared by the author.

Below are the water and sewage operational indicators for private service providers, following the same parameters as the public ones.

Table 3.14

Operational Indicators

Private Service Providers

Qty.	City	1	2	3	4
1	Cachoeira de Itapemirim				
2	Campos dos Goytacazes	7,50		0,0042	
3	Guarà	48,00			
4	Limeira				
5	Mairinque	2,81		0,0028	
6	Niteóri	8,00	129,86	0,0068	3,79
7	Nova Friburgo	6,80	90,00		1,57
8	Petrópolis	8,00			1,50
	Total	4,78	91,53	0,0053	1,00

Source: SNIS 2001, prepared by the author.

By regrouping the average values obtained in the two tables above for the four quality indicators studied so far, the following is obtained:

Table 3.15

Quality indicators

Comparison: Public vs. Private

INDICATORS	PUBLIC	PRIVATE
1 - Average duration of stoppages (Hours/work stoppages)	8,54	4,78
2 - Savings Achieved by Intermittency (Savings/interruption)	96,73	91,53
3 - Incidence of non-standard fecal coliform analysis (percentage)	0,0590	0,0053
4 - Average duration of services performed (Hour/service)	0,91	1,0

Source: SNIS 2001, prepared by the author.

An analysis of Table 3.15 shows that, with regard to the first indicator, 1 - Average Duration of Shutdowns, the index for private service providers is almost 50% better than the public ones, indicating a shorter period of service intermittency, which generates less inconvenience for the affected population.

When we look at the second indicator, 2 - savings affected by intermittencies, we see that the percentages are very close for both public and private utilities, which have an index of over 91%. This could be due to faults in the system, or also to interruptions caused by maintenance on the network, largely due to the archaic physical structure of the supply networks.

In most municipalities, even those with services provided by private companies, the supply network was built in a way that makes it impossible to maintain it without affecting the supply, i.e. they are not sectorized. Thus, when maintenance is needed in a certain part of the municipality, a large part of it is left without supply, which could be avoided if the networks were sectorized, thus reducing the percentage of economies affected by intermittent supply due to maintenance.

The third indicator, 3 - Incidence of fecal coliform analysis outside the standard, is related to public health, which demonstrates the service provider's ability to treat water and supply it to the population with quality and safety, thus following all the potability requirements determined by the Ministry of Health, which regulates that any type of coliform must be absent at the treatment outlets, and in the network, it must be absent in 95% of 100ml samples.

As shown in Table 3.15, this indicator is satisfactory for both service providers, but for private providers, the index is practically zero, and ten times lower than for public providers, demonstrating greater efficiency in chlorination procedures and the reduction of coliforms in the water; therefore, providing a better quality product.

This indicator is also important because it shows, albeit indirectly, whether the service provider has invested in water treatment processes, since a good indicator in this regard is the result of investments in disinfection, chlorination and treatment processes, which today are highly costly.

In the group of quality indicators presented above, there is a big advantage for private service providers, showing from the figures studied that they are more efficient than the public ones. It is worth noting that Tables 3.13 and 3.14 show the absence of indicators for several municipalities. This is true of both public and private service providers.

One reason for this is that it is difficult to measure these indicators, as the service manager may not be interested in this information, not least because of the complexity involved in obtaining it.

The results were in favor of private service providers in practically all groups of indicators, demonstrating their greater efficiency; one factor that is unfavorable to private providers is the higher cost of the tariff, as pointed out in the study.

However, both service providers need new investments, because the structure of water and sewage service providers is old and can no longer meet the demands of city growth or social demands, falling far short of the goal of universalizing services.

With this analysis, one could mention the fact that public service providers are the solution to universalizing services, as well as to having more efficient service providers. The figures do indeed indicate this, but it cannot be categorically stated that privatizing services is the best way to achieve this desired level of efficiency.

3.6 Brazilian cases

This section includes an analysis of Brazilian cases of concessions and public management of basic sanitation services. The study took into account the relevance of some of the cases verified and also the availability of information and innovative experiences in management. Before we start looking at each case, we will present the concessions in force in the country.

Table 3.16

Concessions in the sanitation sector in the Southeast of Brazil

MUNICIPALITIES / STATE	POP.	DEALERSHIP (GROUP)	OBJECT
Araçatuba - SP	157.467	SANEAR (Amafi, Multiservice, Resil, Tejofran)	Sewage
Birigüi - SP	84.016	AQUAPÉROLA (Isratec/ Hidrogesp)	BOT well
Cachoeiro do Itapemirim - RJ	149.820	ND	Full depth
Cajamar - SP	33.707	AGUAS DE CAJAMAR Ltda. (Multiservice, Rek)	Water
Campos - RJ	350.000	AGUAS DO PARAiBA (Developer, Cowan, Queiroz Galvao, Carioca)	Full (water and
Itu - SP	112.939	CAVO ITU (Cavo, Camargo Corrêa)	sewers) Sewers
Jaù - SP	97.354	AGUAS DE MANGADA (Amafi, Multiservice, empr. Portugal)	Water
Jaù - SP		Consórcio C.R. Almeida, SILEC (Itàlia) CIA.	Drains
Jundiai - SP	288.644	SANEAMENTO DE JUNDIAÌ (Augusto Velloso, Coveg, Tejofran)	Drains
Limeira-SP	217.489	AGUAS DE LIMEIRA (Lyonnaise des Eaux, CBPO)	Full
Mairinque - SP	35.000	CIAGUA (Villanova Group)	Full
Marilia - SP	173.841	AGUAS DE MARÌLIA (Hidrogesp)	Water (BOT of well, water main,
Miners of Tietê - SP	9.462	SANECISTE	reservoir) Full
Mirassol - SP	48.312	ND	Full
Niterói - RJ	448.736	AGUAS DE NITERÓI (Cowan, Carioca,	Full
Ourinhos - SP	79.148	Trana, Q. Galvao, Developer)	Water - well

		AGUAS DE ESMERALDA (Hidrogesp, Multiservice)	
Ourinhos-SP		TELAR (Telar Eng.)	
Paranaguà - PR	110.000	AGUAS DE PARANAGUA	deep sewerage Plena
Pereiras - SP	4.850	(Castilho, Carioca, Developer) NOVACON	Full
Petrópolis - RJ	263.838	AGUAS DO IMPERADOR (Cowan, Trana, Q. Galvao, Developer)	Full
Region of Lagos-I RJ (Araruama/Saquarema/Silva Jardim)	200.000	AGUAS DE JUTURNAiBA (Cowan, Trana, Q. Galvâo, Developer, ERCO)	Full
Regiâo dos Lagos-II RJ (C. Frio/ Bùzios/ Arraial/ S.P. Aldeia)		PROLAGOS (Monteiro Aranha/Aguas de Portugal/ PEN)	Full
Ribeirão Preto - SP	450.960	ENVIRONMENT (CH2Mhill/Rek)	Drains
Salto - SP	100.000	SALTO SANECISTE (Saneciste)	Sewage treatment
Sâo Carlos - SP	25.000	ND	Water - BOT
Tuiuti - SP	3.000	RIBEIRÂO DO PÂNTANO - Emp. Saneamento Tuiuti (Novacon)	Full

Source: *apud* Justo 2004, 69 and 70.

As we can see, private sector participation in the basic sanitation sector is still small and restricted. In this respect, Vargas (2003) notes that the small number of private companies in relation to Brazil's market potential is mainly due to the obstacles that still exist from a "legal-legal point of view", especially in terms of ownership.

The cases studied below are: i) Sociedade de Abastecimento de Agua S.A - Sanasa, from the municipality of Campinas - SP; ii) Aguas de Limeira, from the municipality of Limeira - SP; iii) Serviço Autònomo de Agua e Esgoto de Jaboticabal - SAAEJ, from the municipality of Jaboticabal - SP; iv) Departamento de Agua e Esgoto de Ribeirâo Preto - DAERP, from the municipality of Ribeirâo Preto - SP and v) Departamento de Agua e Esgoto de Serrana - DAES, from the municipality of Serrana - SP.

3.6.1 SANASA

Sociedade de Abastecimento de Agua S.A - SANASA is a mixed-capital company whose majority shareholder, 9.9%, is the Campinas City Council. It is responsible for water supply, sewage collection and treatment services in the city of Campinas, State of São Paulo. SANASA is currently considered to be one of the best examples of an efficient and high-quality municipal public service, serving 98% of the population with water supply and 88% with sewage collection[49] , *while* sewage treatment serves approximately 37% of the population.

SANASA is working towards the goal of treating 70% of the sewage in the city of Campinas in a short space of time, and by 2016 having 100% of the municipality's sewage treated, thus complying with its Sewage Treatment Master Plan. An example of this initiative was an agreement signed between SANASA and the municipal governments of Campinas and Valinhos (a neighboring municipality), with the aim of building the Capuava Sewage Treatment Plant[50] , thus enabling Valinhos' sewage to be treated and not discharged into the Pinheiros stream, a tributary of the Atibaia River .[51]

In order to build the Capuava Wastewater Treatment Plant, a Mutual Cooperation Agreement was signed between SANASA and the Valinhos City Council, with the approval of the National Water Agency (ANA), through which SANASA transferred around R$5.5 million in the form of a loan to the Valinhos Water and Sewage Department (DAE), which was only part of the total investment needed - R$11 million, because the Valinhos DAE did not have the resources for such an investment. The loan would only start to be paid back after the final operation of the WWTP.

In addition to this investment, there are others that SANASA has been making, using its own resources as well as financing, loans and transfers from the federal government, with the aim of treating the sewage in the city of Campinas, building ETEs, outfalls and interceptors to collect the sewage and take it to the stations. The goal is huge, as is the population served, more than 1 million, without considering cases like Sumaré, where SANASA sells water to the neighboring municipality, to serve around 200,000 people .[52]

SANASA has been in existence for over 30 years, but since 1887 there had been a private company in Campinas responsible for water and sewage services, Companhia Campineira de Aguas e Esgotos - CCAE, i.e. a 60-year concession, which was later extended for 90 years, but by the 1920s, the system was precarious and in need of investment. CCAE requested authorization from the municipality to adjust its tariffs, but the request was denied, which made the company unviable and it was taken over in 1923. Then came the Water and Sewage Department, which was replaced by the Board of Directors in 1953 and in 1974 SANASA was born.

SANASA is now undoubtedly an important company in many ways, because it is a profitable company, which generates profits, provides excellent quality services to the people of Campinas and, finally, it is a public company, which has overcome the obstacles of politics and preconceptions that public power is inefficient.

In order to reach this level, SANASA promoted extremely efficient, competent management, but above all,

---

49 Source: SANASA.
50 Opened in August 2004.
51 The most important spring for the city of Campinas, as a source of water for treatment and distribution to the population.
52 This was only possible because SANASA produces more water than it consumes.

independent of political management, i.e., political interests over SANASA took precedence over the need and importance of the company for the city of Campinas and its residents.

Until 2000, SANASA faced a serious financial and economic crisis, an example of which was the loss that the company closed that year, R$6 million. After that year, the company's results reversed, and it began to make successive profits. What stands out in this evolution of the company's profitability are points such as: serious and efficient management, technically competent employees, good technical resources, action planning and, in particular and fundamental to the process, the inflationary readjustments and replacements in SANASA's water and sewage tariffs.

The company has promoted measures such as a program to reduce losses in its water supply network, which today are one of the lowest in the country at 23%, as well as implementing a program to reduce the company's costs. However, the tariff adjustments are undoubtedly SANASA's key strengths in the period of financial crisis it was experiencing, as they strengthened the company and allowed for accounting and financial adjustments, which would have been impossible without them.

The next table shows the comparison between SANASA's tariff adjustments and the inflation rate measured by the IPCA (the official government index). It can be seen that, even during the periods in which there were changes in the municipality's political framework, the tariff management policy remained the same, i.e. that of replacing inflation, and in some periods the replacement was even higher.

Table 3.17

Comparison between tariff adjustments and inflation

Year	Tariff Adjustments (in %)	Inflation index (IPCA) -in%
1996	18,50	9,56
1997	9,6	5,22
1998	6,18	1,65
1999	14,20	8,94
2000	10,00	5,97
2001	21,34	7,67
2002	9,48	12,53
Cumulative	130,12	63,90

Source: *apud* Justo 2004, 80.

It is clear from the table above that the most important tool for cleaning up SANASA's accounts were the tariff increases applied. These increases were applied after the public had been made aware of the need for such an attitude, as well as the benefits of such a procedure, so there was popular support for the increase in tariffs, strengthening SANASA and avoiding major criticism and complaints, even though it was an extremely bad action from a political point of view.

This shows that although SANASA is a public company owned by the municipality of Campinas, it has broken with some paradigms, especially regarding its efficiency and quality, making it possible to offer the population quality, efficient services, generate value for the municipality[53] and grow with the city.

53 SANASA is today a company targeted by many private groups investing in the basic sanitation sector, and is owned

All this restructuring of the company was of fundamental importance for operations such as the release of more than R$ 120 million by the federal government for the construction of sewage treatment plants in 2003, and contributed greatly to the company's situation.

3.6.2 Aguas de Limeira

Aguas de Limeira is a private company that operates the total concession of water and sewage services in the city of Limeira, state of São Paulo, and is considered an example of a successful concession in the area of basic sanitation, with positive results.

The concession was made official in 1995, amid controversy, accusations, lawsuits and other events. The bill authorizing the concession, the bidding process for choosing the company and the contract itself have all been criticized, but this is not the subject of this paper.

At the time the concession was granted, the city of Limeira was facing serious difficulties in terms of water supply services, which served 90% of the population, but there were constant rotations and lack of supply, while sewage collection reached 80% of households, but only 2% of everything collected was treated. The municipality of Limeira, like the others, also lacked the financial resources to make the necessary investments. It was in this context that the services were concessioned to Aguas de Limeira, a consortium made up of two companies: the Brazilian group Odebrecht and the French company *Suez Environment*.

It is worth highlighting some aspects of the concession which, in a way, greatly benefited the private company Aguas de Limeira during the concession period and which were of fundamental importance to the company's performance, such as: i) SAAE de Limeira's debts: the concession contract exempted the company Aguas de Limeira from any debts, debts or any other liabilities contracted by SAAE de Limeira before the concession contract was signed; ii) No onerous concession: the concession granted by the municipality of Limeira was not onerous, i.e. the group that formed the company Aguas de Limeira was not obliged to pay a single penny for the concession[54] , only being obliged to present a guarantee of R$ 6 million and insurance for SAAE de Limeira's assets and, iii) Action Planning: the planning, implementation, expansion, operation, maintenance, administration, exploitation and management of basic sanitation services are the responsibility of the concessionaire, taking control of a relevant instrument for urban planning away from the municipality. These are facts that have facilitated and contributed to the concession.

Currently, the company Aguas de Limeira has excellent consumer acceptance rates[5]s, in addition to providing 100% of the population with good quality drinking water and 100% sewage collection, the treatment rate is 70%, with a target of reaching 100% sewage treatment in the municipality of Limeira by 2009.

In order to achieve these figures, Aguas de Limeira invested in various areas. The funds for this came from contributions from the companies, as well as from financing, agreements signed with the federal government and other funding bodies, and of course from its own income.

---

by the municipality of Campinas.

54 Unlike what has happened in recent cases of concessions in the country, in the various sectors of the economy.

A striking fact in this case regarding the performance of Aguas de Limeira refers to the tariff increases that were carried out, which overvalued the tariffs by more than 50% of the previous value practiced by the municipality before the concession, even though the contract stated that the readjustments should be authorized by the city council, in 1996 there was a legal dispute that was ended in 2000, when an agreement was reached between the concessionaire and the municipality, and an readjustment was applied according to the IPCA for the accumulated period without readjustments (1994 - 2000).

In the case of Limeira, the most important fact, apart from the concession itself, was the tariff adjustments applied, even after legal action and political wrangling. An important fact about Limeira is that after the concession, the company began to use new techniques and equipment that had not previously been used by the municipality, due to the bidding law, which prohibited the purchase of imported products, equipment and services, among other restrictions.

These products and equipment were very important for the company's development, such as the use of orthopolyphosphate to improve the appearance of the water distributed, which, after it was implemented by Aguas de Limeira, was used by several other service providers, as well as being regulated by ABNT. In addition to equipment such as new water meters (micro and macro meters), among others.

3.6.3 SAAE of Jaboticabal

The case of Jaboticabal refers to the Municipal Autarchy, Serviço Autônomo de Agua e Esgoto - SAAEJ, created in 1974, a decisive time when the services were handed over to CESBS or taken over by the municipality. SAAEJ has 146 employees serving the entire population of [55]

Jaboticabal (approximately 70,000 inhabitants), as well as its two Districts of Córrego Rico and Lusitânia (the study will be carried out in the settlement of Córrego Rico) in water collection, treatment, reservoir and distribution services, in addition to sewage collection and treatment services in the Districts. SAAEJ also has 27,000 water and sewage connections in the city, as well as its own, as described in the table below:

Table 3.1

SAAEJ data

Locations	Quantity
Reservoirs[56]	26 (13 elevated; 10 semi-underground and 3 supported)
Wells	9 (only one deep)
Drains	3
Booster	3
Lift	4
ETE	2 (Districts of Córrego Rico and Lusitânia)
Administration	1
Post of Service	1

Source: Fenerich, A. P. Carlos, SAAEJ, 2003.

55 Source: Associação Brasileira das Concessionarias de Serviços Pùblicos de Agua e Esgoto - ABCON.
56 SAAEJ's total reserve capacity is 17,358.90 m3.

The municipality has a good tariff structure and values in line with the reality of the services, in addition the suspension of services to non-payers and defaulters is a common occurrence in the municipality and the tariff repositioning until 2004 was always carried out following the IGPM values, after that year, the percentage was not the full IGPM, but only a percentage, which was undoubtedly due to political pressure from the new government that took office in 2005, already showing possible political interference in the management of the municipality.

SAAEJ's annual budget is approximately 7 million reais[57] , which shows a reduced capacity for investment, given the need to maintain services such as: replacing the fleet of hydrometers, maintaining the water and sewage networks, maintaining public buildings, investing in employee training, equipment and vehicles, due to their natural depreciation and, of course, employee payroll costs.

The Jaboticabal City Council has sought out various means, through studies, to make credit lines available for financing, with the aim of investing in environmental sanitation for the decommissioning of the city's old garbage dump, removing families from the garbage, as well as resources for the expansion of water and sewage services in the city.

During this period, practically all lines of credit and even federal government programs ceased, which is why the local government created the Special Investment Fund in 1997, in order to raise funds for the expansion and consolidation of the municipality's water supply and sewage collection and treatment services.

The Fund was created by Municipal Law No. 2,550 of July 18, 1997, which provides for a 15% levy on Jaboticabal's water supply and sewage collection tariffs. Around 70,000 reais a month are collected (2002/2003), which since its creation, i.e. July 1997 to March 2003, has generated an amount of R$4,113,234.43, which has not been corrected by any monetary index. These funds were earmarked for the expansion of the water supply system and sewage collection and treatment in the municipality and its districts.

The Special Investment Fund is overseen by a committee made up of the City Hall, the City Council, ACIJA - the Commercial and Industrial Association of Jaboticabal, OAB - the Brazilian Bar Association, AREA - the Regional Engineering and Architecture Association and SAAEJ, SAAEJ and its technical team are responsible for managing and deciding on the use of resources, and the financial reports on some of the Commission's necessary deliberations are presented monthly.

In 2001, the law creating the Special Investment Fund was amended by Municipal Law No. 2938, of October 15, 2001, changing the fact that the amount of the funds, 15% of SAAEJ's water and sewage tariff, would be earmarked solely and exclusively for the execution of the Works, Services and Acquisitions Necessary for the Implementation of the Issuers and the Sewage Treatment Plant, and also gave a maximum term of four years.

A legal regulation was also carried out in 1997, in view of the pressure from the Public Prosecutor's Office to prevent the *"in natura"* discharge of sewage into the city's streams; thus, in 1998, Municipal Law No. 2.633 of March 23, 1.998, which authorized the Jaboticabal City Council to sign a Commitment Agreement with the

57 Source: Santiago, R., SAAEJ, 2003, data for the year 2003.

São Paulo State Public Prosecutor's Office, aimed at repairing environmental damage resulting from the *"in natura"* discharge of domestic sewage, within a period of seventy-two months, extendable for a further twenty-four months, starting in March 1.998.

This has made it possible to invest in various areas of sanitation, with a policy for: water, sewage, solid waste, drainage, green areas and environmental education, with the aim of providing the population with water, collecting sewage and treating it. The municipality currently supplies 100% of its residents with water and has 100% sewage collection. Treatment has the following characteristics: the districts of Córrego Rico and Lusitânia have 100% of their sewage treated, using small sewage treatment plants, while the municipality of Jaboticabal finished building the city's sewage treatment plant in 2005, but to date it has not been put into operation, and SAAEJ has not provided any justification for this.

The solution found by the municipality is very interesting because, due to the fact that it doesn't have the capacity to generate its own resources, nor the possibility of using a line of credit from the government or any other financing organization, the municipality has found a way to solve its problem.

The creation of the fund is characterized by a surcharge on water and sewage bills, in addition to having greater oversight by civil society, through the council, so two facts are evident: i) an increase in the value of water bills, meaning that the government needed a financial contribution from the town's residents to finance the work; ii) participation and supervision by society: with greater participation by organized civil society, it was possible for the resources obtained from the fund to be properly applied and good results obtained.

3.6.4 Ribeirao Preto

The case of the municipality of Ribeirâo Preto, state of São Paulo, is a partial concession of the sewage treatment service. During the concession period (1994), the municipality treated only 6% of its sewage, since the Ribeirâo Preto Water and Sewage Department (DAERP), the authority responsible for basic sanitation services in the city, and the city council did not have the resources to carry out sewage treatment services.

The concession took place amid electoral disputes and questions, but on September 28, 1995 the concession contract was signed between Ribeirão Preto City Hall and Ambient, a company formed by a consortium made up of REK and CH2M Hill International. The consortium had an agreement with BNDES, which would provide loans for the initial investments needed, with the loan limit being 65% and the company having to make a contribution of 25% of the total value of the investments.

The company would be paid through the Basic Sewage Treatment Tariff (TBTE), and DAERP would only start charging after Ambient had started treating the sewage. The project foresaw the construction of three sewage treatment plants to achieve 100% of the sewage treated in Ribeirão Preto.

There was a requirement for Ambient to maintain guarantee insurance and also a bond of R$1.5 million, in addition to the free and automatic reversion of the assets that make up the concession to DAERP at the end of the 20-year concession.

As the BNDES did not release the promised funds, the works and the timetable were not carried out, and by

1997 only 6% of the project had been completed. After several disputes, fines and lawsuits, both the municipality and Ambient reached an agreement in which the American company CH2M Hill was replaced by the Spanish company Inima, tariffs were readjusted, concessions were granted on both sides and in 2000 the BNDES released R$35 million for the necessary investments.

For the concessionaire, the risks of default are practically nil, as DAERP must pay Ambient an amount for the volume of sewage treated, but DAERP assumes the risk of defaults on water and sewage bills, thus reducing the business risks for the concessionaire.

In this case, we can see that there was also a readjustment of tariffs, as well as the financing of the business with public money, in this case through financing from the BNDES.

3.6.5 Serrana

The municipality of Serrana, in the state of São Paulo, is located in the region of Ribeirão Preto (20 km), and is a small municipality. In 1998, discussions began for a concession, and in 1998 a contract was signed for the concession of water and sewage services to the company Novacon Engenharia de Concessoes Ltda and the assignee Empresa Bela Fonte Ltda. The reasons for the concession were the difficulty in supplying the population with water services and the irregularity of the supply, as well as the lack of sewage treatment and low sewage collection.

The concessionaire had great difficulty in operating the system and solving the water supply problem, until the new mayor terminated the contract and the entire concession on January 28, 2001, by means of a municipal decree. There was a lot of popular support for this, due to dissatisfaction with the services and the constant lack of water in the municipality. As a result, the water and sewage services were taken over by the municipality, and carried out by the Serrana Water and Sewage Department - DAES.

Currently, 100% of the city has a water supply and, most importantly, the system is regular, meaning that there is hardly any shortage of water or suspension of services due to technical problems. As far as sewage is concerned, several works have been started, such as outfalls, interceptors and others, as well as the completion of the project to build a sewage treatment plant, which will enable 100% of the city's sewage to be treated, in addition to defining the area for its implementation. Until 2000, there were no prospects.

The main factors contributing to the technical improvement of services were investments by the municipality, as well as major investments made by the federal government through agreements from 2005 onwards[58] , in addition, various administrative and technical measures were implemented, such as: the suspension of the water supply for debtors, outsourcing of some services, re-registration of the network, a preventive maintenance program for the city's wells, and re-dimensioning of the entire system, in addition to carrying out various projects to raise funds from ministries such as health and cities, as well as the necessary tariff readjustments, at least equal to inflation for the period.

All these measures together made it possible to make the investments made, as well as to improve DAES's

58 Source: DAES, 2006.

performance. This was of course reflected in revenues, which practically doubled from 2000 to 2005, to R$1.7 million .[59]

3.7 Chapter Conclusions

When comparing the performance of public and private basic sanitation providers in the south-east of the country, we can see that their performances are similar, and the differences are not great. However, in several indices and indicators, the private providers seem to be more technically efficient, But this is justified by a series of reasons and periods of time, such as the way in which the public authorities are contracted, the way in which private companies are managed to make a profit and other factors, but the most obvious of all is the tariff charged. As seen in the studies, private sanitation service providers have higher tariffs than public ones.

Of course, some caveats must be made, given the sample that was used, with the exclusion of Sabesp, but the information is nonetheless truthful and interesting for analysis.

In addition, a fact that predominated in practically all the cases studied, in which there was a private concession or privatization of services or even in those that gained efficiency over time, was the increase in the tariff charged by the providers, or even the creation of new tariffs, as in the case of Jaboticabal. This leads to the conclusion that the providers studied perform well not only because of their technical capacity, open lines of credit and financing, but also because of the sanitation service provider's austere tax management, with a commitment to charging tariffs and rates that are realistic for the business.

This view is often criticized by various segments of society, which assume that the government should subsidize the costs of public services; this thinking is true to a certain extent, but this subsidy must not make a service provider unviable in such a way that all its services are compromised and many individuals suffer, with a lack of water, sewage collection, treatment, etc.

Therefore, part of any restructuring of a basic sanitation service, whether provided by the public or private sector, is the need to adjust the tariff structure so that it pays for all the services provided. Of course, cross-subsidization can and should occur so as not to exclude consumers, as this is an essential public good, but it must be done carefully and fairly.

59 Source: DAES, 2006.

FINAL CONSIDERATIONS

The difficulties faced by the basic sanitation sector are manifold: low coverage, poor quality of services, problems of political interference, low government investment (due to lack of financial availability), etc. All this serves as an argument for the government to seek alternative ways of financing the sector.

The federal government has been trying to promote the entry of private sector resources into the infrastructure sectors, which is why some actions have been carried out, such as: the implementation of the concessions law, which was widely used during the privatization period of the FHC government[60] and recently the Public-Private Partnership, which in the government's view will be a possible matrix for generating new private investment in the public sector, especially in infrastructure.

So far, however, there hasn't been much leverage of private sector investment, nor are the PPPs fulfilling the role for which they were implemented by the Lula government. There are various other forms of private capital participation in the public sector, as we have seen: (i) management and outsourcing contracts; (ii) *leasing;* (iii) concessions (with the models: BOT, BOOT and BOT) and (iv) privatizations, in addition, there is the participation of the National Bank for Social Economic Development - BNDES, Caixa Econômica Federal in the sense of assisting technically and even financially the entry of these new players into the public sector.

Despite all this, the sector still doesn't have much prospect of a solution, because as has already been pointed out, the government doesn't have the financial resources to invest in sanitation and the private sector is still a little wary of entering sanitation services, for a number of reasons, including lack of regulation, especially the lack of legislation in the sector, doubts about the return on investment, contractual security in the execution of concession contracts and also the fact that the sanitation sector is not on the government's urgent agenda.

The international experiences of private capital participation in the public sector, in the case of this paper England and France, show us some common characteristics: i) a large state participation in the structuring of the sector, either by promoting actions to organize it in order to promote privatization, or to clean up the accounts of the providers in order to make it attractive to private capital and, ii) financing of the companies with tariff resources, which resulted in large tariff increases in the cases studied.

The Brazilian cases studied have shown that in all of them the tariff factor was essential for the resumption of services, as well as for the improvement in quality and efficiency of the basic sanitation service provider.

This becomes clear when we look at the case of Limeira, which was privatized and experienced a major price increase in tariffs. The same happened in Campinas with Sanasa, which needed to increase its tariffs in order to solve its alarming financial problems. The city of Serrana, soon after the municipality took over the services of the private provider, also needed to increase its tariffs, as well as start suspending the supply of water to debtors so that it could obtain resources to guarantee the supply, which was very precarious and inefficient.

The municipalities of Ribeirao Preto and Jaboticabal, in a way, also applied adjustments to their water and sewage bills to make investments viable, with the aim of implementing sewage treatment services. In the case

60 Energy, transport and telecommunications sectors.

of Ribeirao Preto, this is a concession, in which Ambient (the concessionaire) receives a fee per cubic meter of treated sewage, and consumers pay a fee for this service after the start of the activities. In Jaboticabal, the services are provided by a municipal authority, which has set up an investment fund that taxes water and sewage bills at 15%, meaning that consumers have to pay 15% more on their water and sewage bills to make the municipality's sewage treatment viable.

As seen in both the international cases and the relatively successful national ones (even those in which there is no private capital participation), the tariff increase was a fundamental and necessary factor to make investments in the sanitation sector viable, without which the situation of these providers would hardly have been any different.

Another fact that should be analyzed is the comparison of performance between public and private service providers in southeastern Brazil. After comparing the data, it can be seen that when the private sector participates, tariff increases are a common occurrence.

Undoubtedly, the participation of the private sector in urban infrastructure services, especially basic sanitation, is a way out that the country is moving towards. There are many problems with this, but the maintenance of services with the public sector has not shown great improvements: changes to legislation, limits on indebtedness, excessive credit contingency for the sector, fiscal adjustment and more. Privatization, therefore, is not the only solution for the sector, as it alone does not guarantee a good result.

One point worth highlighting is that, as the study found, the best efficiency of the basic sanitation service provider is related to tariff increases, whether the provider is public or private, which is a problem for the poor, low-income population, as they run the risk of limiting their access to services. In this way, some mechanism must be found that does not make it impossible for this population to access the services, whether through cross-subsidization, social tariffs or others.

BIBLIOGRAPHY

ABCON - Associação Brasileira das Concessionârias Privadas de Serviços Pùblicos de Agua e Esgoto. Available at: http://www.abcon.com.br. Accessed on: April 20, 2006.

ABICALIL, M. Os serviços de sanamento no Brasil: transformar para universalizar. Rio de Janeiro: FASE/Caixa Econômica Federal, 1998.

ALÉM, A.; GIAMBIAGI, F.. *Finanças Pùblicas - Teoria e Pràtica no Brasil.* 2.ed. Rio de Janeiro: Campus, 2001.

ALVES, F. *Saneamento Ambiental.* Sginus 2004. ano XIV, n° 107 jul./ago. 2004.

ARAÙJO, R. "Regulaçâo da Prestaçâo de Serviços de Saneamento Bàsico - Abastecimento de água e Esgotamento Sanitáriario". *Infra-Estrutura Perspectivas e Reorganizaçao - Saneamento,* Brasilia: IPEA, 1999.

ARRETCHE, M.T.S. *O Fundo de Garantia por Tempo de Serviço e as Politicas Habitacionais e de Saneamento Bàsico;* IPEA - Ministério da Fazenda.

ARRETCHE, M.T.S. *Politica Nacional de Saneamento: A Reestruturaçao das Companhias Estaduais;* IESP/FUNDAP, mimeo, 1993.

ASCHAUER, D. A. "Is public expenditure productive?" *Journal of Monetary Economics,* Holland, n° 23: p. 177-2000, 1989.

ASSEMAE - National Association of Municipal Sanitation Services. *Newsletter,* Brasilia, ASSEMAE n° 108: jul. 2003.

. *Newsletter,* Brasilia, No. 112: ASSEMAE, June 2004.

BAGGIO, M.. "Water Losses: A New Approach to Avoiding Them". *ANAIS da ASSEMAE - 3ª Exposiçao de Experiências Municipais em Saneamento*, Vitória: ASSEMAE, jun. 1998.

BNDES (1994). National Privatization Program. Rio de Janeiro, 1994.

BNDES. "Basic Sanitation Services - Service Levels". *Informe Infra-Estrutura*, Rio de Janeiro, BNDES No. 8: Mar. 1996a.

"Sanitation Sector - Paths Adopted". *Informe Infra-Estrutura*, Rio de Janeiro, BNDES n° 20: jun. 1998b.

"Sanitation: The Goal is Efficiency". *Informe Infra-Estrutura*, Rio de Janeiro, BNDES n° 23: jun. 1998c.

"Photograph of Private Participation in the Sanitation Sector". *Informe Infra-Estrutura*, Rio de Janeiro, BNDES No. 32: June 1999.

CALMON, K. et alii. "Saneamento: As Transformações Estruturais em Curso na Açâo Governamental". *Infra-Estrutura Perspectivas e Reorganizaçao - Saneamento*, Brasilia: IPEA - Instituto de Pesquisa Econômica Aplicada, 1999.

CASAGRANDE NETO, H. *Abertura do capital de empresas no Brasil.* 2ª ed. Sâo Paulo: Atlas, 2000.

DECREE NO. 13/2001. Provides for the assumption of water and sewage services by the Serrana City Council, March 2001. Serrana City Hall - SP.

FERNANDEZ, C. The management of basic sanitation services in Brazil. *Scripta Nova. Electronic journal of geography and social sciences. Barcelona:* University of Barcelona, August 1, 2005, vol. IX, no. 194 (73). <http://www.ub.es/geocrit/sn/sn-194-73.htm> . Accessed on November 10, 2006.

GUASCH, L. *Garanting and Revegotiating - Infrastructure Concession - Doing it Right.* 1ª ed. Washington, D.C.: The World Bank, 2003.

HASENCLEVER, L; KUPFLER, D. *Economia Industrial: Fundamentos Teóricos e Pràticas no Brasil.* 1st ed. Rio de Janeiro: Campus, 2002.

IPEA - Institute for Applied Economic Research - Ministry of Planning and Budget. *Infra-Estrutura - Perspectivas de Reorganizaçao - Financiamento, p. 140,* 1998.

JOURAVLEV, A. *Water utility regulation: issues and options for Latin America and the Caribbean.* Economic Commission for Latin America and the Caribbean (ECLAC), Oct. 2000.

JUSTO, M. "A comparative analysis of public and private management in the financing of the basic sanitation sector". *Anais 8ª Exposiçâo de Experiências Municipais em Saneamento - 34ª Assembléia da Nacional da ASSEMAE*, Rio de Janeiro: ASSEMAE, May 2003.

JUSTO, M. *Financiamento do Saneamento Bàsico no Brasil" - A Comparative Analysis of Public and Private Management.* 2004. 167f. Master's Thesis in Economic Development, Space and Environment - in the area of concentration in Regional and Urban Economics - State University of Campinas - Campinas Institute, Campinas.

LAW NO. 10.735. Incentive Program for the Implementation of Social Interest Projects (PIPS). Sept. 2003. Available at< http://wwwt.senado.gov.br>. Accessed on: October 15, 2006.

LAW NO. 8.666. Bidding Law. June 1993. Available at< http://wwwt.senado.gov.br>. Accessed on: June 10, 2006.

LAW NO. 8.987. Concessions Law, Feb. 1995. Available at< http://wwwt.senado.gov.br>. Accessed on: March 10, 2006.

COMPLEMENTARY LAW N° 4.320. Public Accounting Law, Mar. 1964. Available at< http://wwwt.senado.gov.br>. Accessed on: June 10, 2006.

COMPLEMENTARY LAW No. 67/98, of February 25, 1998 - Provides for the granting of public water and sewage services in the municipality of Serrana, Serrana City Hall - SP.

COMPLEMENTARY LAW No. 101, of May 4, 2000 - Establishes public finance rules aimed at responsibility in fiscal management and makes other provisions, Presidency of the Republic.

LOPES, L. et alii. "Human Resources Management at the SAAEJ WTP*". Annals of ASSEMAE - 7th Exhibition of Municipal Experiences in Sanitation,* Caxias do Sul: ASSEMAE, May 2004.

MENDONÇA, M. and MOTTA, R. "Health and Sanitation in Brazil"; *Text for Discussion.* IPEA n° 1081, Brasilia, 2005.

MINISTRY OF CITIES - NATIONAL SANITATION INFORMATION SYSTEM. *Diagnosis of Water and Sewerage Services - Historical Series 1995 - 2001,* 2002.

MINISTRY OF HEALTH. *Ordinance No 1.469, of December 29, 2000 - Establishes the procedures and responsibilities relating to the control and surveillance of the quality of water for human consumption and its standard of potability, and makes other provisions.*

MONTENEGRO, L. "Sanasa stands out among the municipalities".*Revista Saneamento Ambiental,* Sao Paulo, 109: 30-35, Sept./Oct. 2004.

MONTENEGRO, L. "Investimentos em Saneamentos".*ASSEMAE*, Brasilia, 2000.

MORAES, J. "O processo da qualidade no SEMASA - A Certificaçao ISO 9002 e a Implantaçao de uma Política de Gestao". *Anais da ASSEMAE - 3ª Expos^ão de Experiências Municipais em Saneamento*, Vitória: ASSEMAE, jun. 1998.

MOREIRA, T. Saneamento bàsico: desafios e oportunidades. s.d. Available at:< http://www.bndes.gov.br> . Accessed on: September 10, 2004.

. "A hora e a vez do saneamento. *Revista do BNDES*, Rio de Janeiro, No. 10: Dec. 1998.

. "BNDES 50 anos - Histórias Setoriais: a Infra-Estrutura Urbana". *BNDES Magazine,* Rio de Janeiro: Dec. 2002.

MUSGRAVE, R.; MUSGRAVE, P. *Public Finance*. Translated by Carlos A. P. Braga; Clàudia C. C. Eris. Sao Paulo: University of Sao Paulo, 1980.

PARLATORE, A. Privatization of the sanitation sector in Brazil. In: BNDES. *Privatization in Brazil: The case of public utilities*, Rio de Janeiro. 2000.

PINDYCK, R.; RUBINFELD, D. *Microeconomics.* 4ª ed. Trad. de Luis Felipe Cozac; Marcos Teixeira de Barros; Mauro Teixeira Pinto (trad.). Sao Paulo: Makron Books, 1999.

PREFEITURA MUNICIPAL DE JABOTICABAL, Relatório da Administraçao Pùblica Indireta, Serviço Autònomo de Agua e Esgoto de Jaboticabal, 2006.

REIS, R. "Aguas de Limeira: *saneando* conceitos".*Revista Saneamento Ambiental,* Sao Paulo, 109: 36-41, Sep/Oct 2004.

REZENDE, F. *Finanças Pùblicas.* 2. ed. Sao Paulo: Atlas, 2001.

RIANI, F. *Economia do Setor Pùblico: Uma Abordagem Introdutória.* 4th ed. Sao Paulo: Atlas, 2002.

SAIANI, C. C. S. Alternativas de Financiamento para Investimentos em Saneamento Bàsico no Brasil, Fapesp Scientific Initiation Report, 2004.

SANCHES, O. The privatization of sanitation. Text for discussion. Sao Paulo, 2000.

SHUGART, C. *Regulation - by - Contract and Municipal Services: The Problem of Contractual Incompleteness.* Cambridge, Massachusetts: Harvard University, 1998.

TIROLE, J. *The Theory of Industrial Organization* Massachusettts: MIT, 1988.

TONETO JUNIOR, Rudinei. *The Current Situation of Basic Sanitation in Brazil: Problems and Prospects.* 2004. 323f. Thesis in Economics - Ribeirao Preto School of Economics, Administration and Accounting, Ribeirao Preto.

TUROLLA, F. "Basic sanitation policy: recent advances and future public policy options". *Textos para Discussão do IPEA*, Brasilia, n° 922: p. 1-26, dec. 2002.

VASCONCELLOS, μ.; OLIVEIRA, R. *Manual de Microeconomia. 2.* ed. Sao Paulo: Atlas, 2000.

VASCONCELLOS, M., TONETO JUNIOR, R.; GREMAUD, A.. *Brazilian and Contemporary Economics.* 4ª ed. Sao Paulo: Atlas, 2002.

VARGAS, M. C. and LIMA, R. F. Water supply and sanitation in Brazilian cities: risks and opportunities of private involvement in service provision. In: *I Encontro Nacional em Pesquisa Ambiental e Sociedade, Aguas questoes sociais, politico- institutional e territoriais*, Campinas, Dec. 2003.

Printed by Books on Demand GmbH, Norderstedt / Germany